Y0-AAY-448

Relaxin and Related Peptides

ADVANCES IN EXPERIMENTAL MEDICINE AND BIOLOGY

Recent Volumes in this Series

Volume 603
THE GENUS YERSINIA: FROM GENOMICS TO FUNCTION
Edited by Robert D. Perry and Jacqueline D. Fetherson

Volume 604
ADVANCES IN MOLECULAR ONCOLOGY
Edited by Fabrizio d'Adda di Gagagna, Susanna Chiocca, Fraser McBlane and Ugo Cavallaro

Volume 605
INTEGRATION IN RESPIRATORY CONTROL: FROM GENES TO SYSTEMS
Edited by Marc Poulin and Richard Wilson

Volume 606
BIOACTIVE COMPONENTS OF MILK
Edited by Zsuzsanna Bösze

Volume 607
EUKARYOTIC MEMBRANES AND CYTOSKELETON: ORIGINS AND EVOLUTION
Edited by Gáspár Jékely

Volume 608
BREAST CANCER CHEMOSENSITIVITY
Edited by Dihua Yu and Mien-Chie Hung

Volume 609
HOT TOPICS IN INFECTION AND IMMUNITY IN CHILDREN VI
Edited by Adam Finn and Andrew J. Pollard

Volume 610
TARGET THERAPIES IN CANCER
Edited by Francesco Colotta and Alberto Mantovani

Volume 611
PEPTIDES FOR YOUTH
Edited by Susan Del Valle, Emanuel Escher, and William D. Lubell

Volume 612
RELAXIN AND RELATED PETIDES
Edited by Alexander I. Agoulnik

Relaxin and Related Peptides

Edited by
Alexander I. Agoulnik, Ph.D.
Department of Obstetrics and Gynecology
Baylor College of Medicine, Houston, Texas, U.S.A.

Springer Science+Business Media
Landes Bioscience

Springer Science+Business Media
Landes Bioscience

Printed in the U.S.A.

Springer Science+Business Media, LLC, 233 Spring Street, New York, New York 10013, U.S.A.
http://www.springer.com

Please address all inquiries to the Publisher:
Landes Bioscience, 1002 West Avenue, 2nd Floor, Austin, Texas 78701, U.S.A.
Phone: 512/ 637 6050; FAX: 512/ 637 6079
http://www.landesbioscience.com

Relaxin and Related Peptides edited by Alexander I. Agoulnik, Landes Bioscience / Springer Science+Business Media, LLC dual imprint / Springer series: Advances in Experimental Medicine and Biology

ISBN: 978-0-387-74670-8

Library of Congress Cataloging-in-Publication Data

Relaxin and related peptides / edited by Alexander I. Agoulnik.
p. ; cm. -- (Advances in experimental medicine and biology ; v. 612)
Includes bibliographical references and index.
ISBN 978-0-387-74670-8
1. Relaxin. I. Agoulnik, Alexander I. II. Series.
[DNLM: 1. Relaxin--physiology. 2. Receptors, G-Protein-Coupled--physiology. 3. Receptors, Peptide--physiology. W1 AD559 v.612 2007 / WP 530 R3817 2007]
QP572.R46R45 2007
573.6'654--dc22

2007035913

About the Editor...

ALEXANDER I. AGOULNIK is an Associate Professor at the Department of Obstetrics and Gynecology, Baylor College of Medicine in Houston, Texas, U.S.A. His main research interests include male and female reproductive tract development, germ cell biology, tumorigenesis, and the role of relaxin peptides in these processes. Alexander Agoulnik received his Ph.D. from the Russian Academy of Science at the Institute of Cytology and Genetics, Novosibirsk, Russia, and did his postdoctoral training at the Max Planck Institute for Biology, Tübingen, Germany.

PREFACE

When Dr. Aaron Hsueh and his colleagues reported the discovery of the relaxin receptors in *Science* in 2002, the editorial written by Dr. Richard Ivell was entitled "This hormone has been relaxin' too long!" Indeed, despite the fact that relaxin was the first peptide hormone discovered more than 80 years ago in 1926, the study of this peptide was carried out only in a limited number of laboratories until now. Relaxin research was likely muted by the failure to identify the relaxin receptor, the variety of biological and often relatively mild effects of this hormone, apparent differences in expression and functions between species, and somewhat ambiguous results of early clinical trials. New approaches to relaxin biology were needed. Not surprisingly the turning point in relaxin research was directly linked to progress in genomic research. The completion of human, mouse and other genome sequences and the production of transgenic mouse mutants directly led to identification of new members of the relaxin family peptides and subsequently to the identification of their cognate receptors. Both the established as well as the novel functions of the relaxin peptides are now under increased scrutiny by pharmacologists, experimental biologists, and clinical scientists.

This book is intended to show the established paradigms and the contradictions, the most recent findings and the future directions in the field of relaxin research. The vast array of diverse topics are highlighted, ranging from the evolution of relaxin family peptides and receptors to their cell signaling, from the role of relaxin in reproduction to the newly discovered functions in cancer progression and in the nervous system. The book includes chapters written by well-known experts in the relaxin field who are actively involved in shaping this rapidly evolving science. Each chapter represents a separate review of a particular area, and therefore some repetition was unavoidable. It is not surprising that a careful reader might discover also different points of view, opinions and sometime disagreements. Even the nomenclature of the relaxin peptides and their receptors is still being debated. It is our hope that such a discussion will usher in the new understanding of relaxin functions.

We currently stand at a critical juncture of scientific research—a vast array of new tools are available for the study of biology. There is no doubt that the application of new in vitro technologies, systematic analysis of gene expression and interactions, computer modeling, and new model organisms will further promote our research field. The various therapeutic and diagnostic applications of relaxin family peptides are intriguing and very promising. The described advances portend the great possibilities. Much is to be done and the field is certainly not relaxin' anymore!

The book is addressed to a wide audience of researchers and students in various areas of basic biology and medicine. Finally, I would like to acknowledge the critical input of each of the authors to the research of relaxin and related peptides and their earnest contributions to the book.

Alexander I. Agoulnik, Ph.D.

PARTICIPANTS

Alexander I. Agoulnik
Department of Obstetrics
and Gynecology
Baylor College of Medicine
Houston, Texas
U.S.A.

Ravinder Anand-Ivell
Research Centre for Reproductive
Health and School of Molecular
and Biomedical Science
University of Adelaide
Adelaide, South Australia
Australia

Ross A.D. Bathgate
Howard Florey Institute
University of Melbourne
Parkville, Victoria
Australia

Joanna Bialek
Universitätsklinik und Poliklinik
für Allgemein Viszeral- und
Gefäßchirurgie
Martin-Luther-Universität
Halle-Wittenberg
Halle/Saale
Germany

Erika E. Büllesbach
Department of Biochemistry
and Molecular Biology
Medical University of South Carolina
Charleston, South Carolina
U.S.A.

Kirk P. Conrad
Department of Physiology
and Functional Genomics
University of Florida College
of Medicine
Gainesville, Florida
U.S.A.

Andrew L. Gundlach
Department of Anatomy
and Cell Biology
Centre for Neuroscience
Howard Florey Institute
The University of Melbourne
Parkville, Victoria
Australia

Kee Heng
Research Centre for Reproductive
Health and School of Molecular
and Biomedical Science
University of Adelaide
Adelaide, South Australia
Australia

Cuong Hoang-Vu
Universitätsklinik und Poliklinik für
Allgemein Viszeral-
und Gefäßchirurgie
Martin-Luther-Universität
Halle-Wittenberg
Halle/Saale
Germany

Sabine Hombach-Klonisch
Department of Human Anatomy and Cell Science
University of Manitoba
Winnipeg, Manitoba
Canada

Richard Ivell
Research Centre for Reproductive Health
School of Molecular and Biomedical Science
University of Adelaide
Adelaide, South Australia
Australia

Arundhathi Jeyabalan
Department of Obstetrics, Gynecology and Reproductive Sciences
University of Pittsburgh School of Medicine
Magee-Womens Research Institute
Pittsburgh, Pennsylvania
U.S.A.

Thomas Klonisch
Department of Human Anatomy and Cell Science
University of Manitoba
Winnipeg, Manitoba
Canada

Edna D. Lekgabe
Howard Florey Institute of Experimental Physiology
Medicine and Department of Biochemistry and Molecular Biology
University of Melbourne
Parkville, Victoria
Australia

Sherie Ma
Howard Florey Institute
The University of Melbourne
Parkville, Victoria
Australia

Ishanee Mookerjee
Howard Florey Institute of Experimental Physiology
Medicine and Department of Biochemistry and Molecular Biology
University of Melbourne
Parkville, Victoria
Australia

Jaqueline Novak
Department of Biology
Walsh University
Canton, Ohio
U.S.A.

Laura J. Parry
Department of Zoology
University of Melbourne
Parkville, Victoria
Australia

Yvonne Radestock
Universitätsklinik und Poliklinik für Allgemein Viszeral- und Gefäßchirurgie
Martin-Luther-Universität Halle-Wittenberg
Halle/Saale
Germany

Chrishan S. Samuel
Howard Florey Institute of Experimental Physiology
Medicine and Department of Biochemistry and Molecular Biology
The University of Melbourne
Parkville, Victoria
Australia

Christian Schwabe
Department of Biochemistry and Molecular Biology
Medical University of South Carolina
Charleston, South Carolina
U.S.A.

O. David Sherwood
Department of Molecular
and Integrative Physiology
University of Illinois
Urbana, Illinois
U.S.A

Sanjeev G. Shroff
Department of Bioengineering
University of Pittsburgh
Pittsburgh, Pennsylvania
U.S.A.

Lenka A. Vodstrcil
Department of Zoology
University of Melbourne
Parkville, Victoria
Australia

Tracey N. Wilkinson
Howard Florey Institute
University of Melbourne
Parkville, Victoria
Australia

CONTENTS

FOREWORD

A historical perspective of relaxin begins with Frederick Lee Hisaw. Hisaw was born in Ozark country in a place called Jolly, MO, in 1891. Throughout his distinguished scientific career, Hisaw enjoyed telling humorous stories about rural Ozark life and was said to excel at this art. After receiving a Master's degree from the University of Missouri in 1916, Hisaw took a position at Kansas State University Experimental Station, where he was put in charge of a program aimed at eradication of the plains pocket gopher, an agricultural pest. In the course of collecting 1200 of these animals, he became aware of marked differences in the pelvic girdle structure between adult females and males. During summers, Hisaw conducted studies toward the Ph.D. degree in the Department of Zoology at the University of Wisconsin by examining the regulation of the resorption of the pubic bones that occurred at puberty in female plains pocket gophers. After discovering that ovarian hormones were involved, he was awarded both the Ph.D. degree and a faculty position in the Department of Zoology in 1924. Hisaw shifted his experimental attention from the plains pocket gopher, which is not found in central Wisconsin, to the guinea pig, which had long been known to exhibit a loosening of the connective tissue at the symphyseal union near parturition. Hisaw suspected this phenomenon was attributable to internal secretions. In 1926 Hisaw discovered relaxin when he found that the injection of serum from pregnant guinea pigs or rabbits into estrogen-dominated guinea pigs induced a relaxation of the pelvic ligaments.[1] In 1930 Hisaw and two of his graduate students extracted the active substance from pig corpora lutea, postulated that it was a peptide-like hormone, and named it relaxin.[2] Further interest in relaxin was minimal until after the Great Depression of the 1930s and World War II in the 1940s when the first surge in relaxin research took place (Fig. 1). From the late 1940s through the early 1960s, impure preparations of relaxin were reported to have numerous effects on the reproductive tract in non-pregnant animals. The pioneering discoveries that relaxin promotes cervical growth, inhibits spontaneous contractility of the uterus, and promotes growth of the uterus predicted physiological roles that were later established for endogenous relaxin in pregnant animals.[3] During the late 1950s and early 1960s, the pharmaceutical company Warner-Chilcott was at the forefront of relaxin research. The company prepared large amounts of a partially purified preparation of porcine relaxin called Releasin that was used for basic research by scientists at Warner-Chilcott and elsewhere.

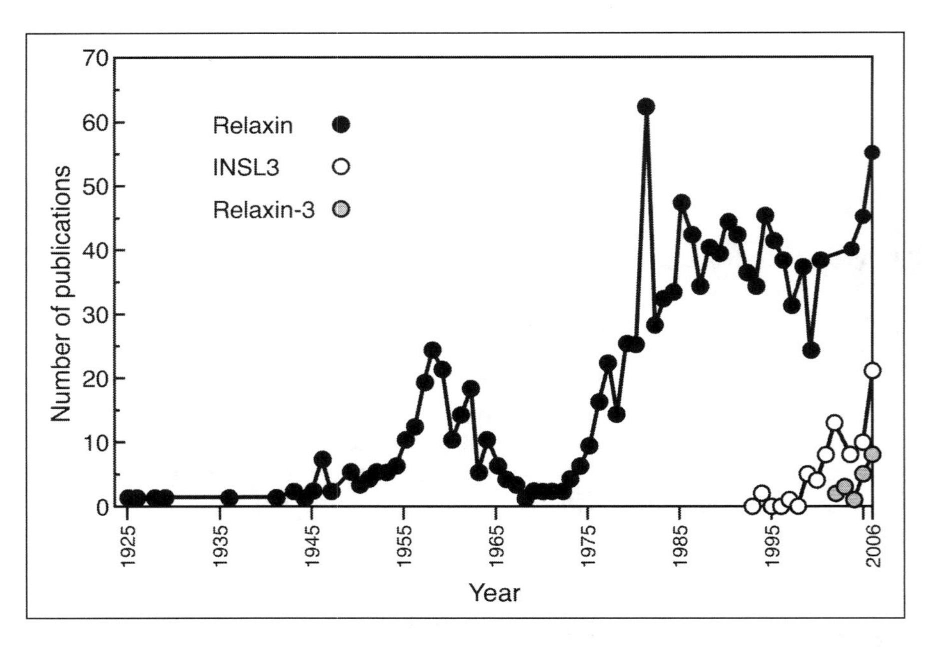

Figure 1. Number of publications per year with the word relaxin, INSL3 (Ley-I-L, RLF) or relaxin-3 in the title. Information is from OLDMEDLINE and PubMed. The approximately 70 publications that appeared in 2005 in the proceedings of the Fourth International Conference on Relaxin and Related Peptides (Ann. NY Acad Sci, Vol. 1041) are not included in the figure.

Releasin was also examined as a potential therapeutic agent for three human clinical problems. It was postulated that by softening the cervix, relaxin would reduce the duration of labor; by inhibiting uterine contractility, relaxin would prevent premature labor; and by increasing skin elasticity, relaxin would have beneficial effects in patients with progressive systemic sclerosis (scleroderma). Clinical efforts with impure porcine relaxin were not sustained beyond the early 1960s for several reasons, including lack of consistent effectiveness, safety problems and the time and expense associated with meeting new and more stringent federal regulatory requirements

There was nearly no research interest in relaxin for about a 10-year period. Then in the late 1960s, the Swiss pharmaceutical company CIBA decided to isolate and characterize porcine relaxin at its facility in Summit, NJ. Those efforts took advantage of new techniques for isolating proteins and a specific and objective mouse interpubic ligament relaxin bioassay developed by Steinetz, Kroc and coworkers at Warner-Chilcott.[4] In 1974, while employed at CIBA-Geigy, Sherwood and O'Byrne published a technique for isolating porcine relaxin,[5] and a few years later techniques for isolating relaxin from several other species, including the rat,[6,7] horse[8] and dog,[9] were reported by Sherwood and others: John and coworkers at the Howard Florey Institute in Melbourne and Stewart and colleagues at the University of California in Davis. These pure hormone preparations led to a resurgence of interest in relaxin in the late 1970s, and numerous investigators who devoted much of their research careers to this hormone sustained this resurgence. Scientists who come to mind are L. Anderson, T. Bani Sacchi, G. Bryant-Greenwood, E. Bullesbach, P. Fields, L. Goldsmith, D. Porter, B. Sanborn, C. Schwabe, D. Sherwood, B. Steinetz, D. Stewart, A. Summerlee, G. Tregear and E. Unemori. Especially noteworthy is Bernard G. Steinetz, who began conducting relaxin research in the mid-1950s, and 50 years later continues to have a relaxin research program. Over the last approximately 30 years, these investigators used highly purified relaxin to conduct rigorous chemical and physiological studies of relaxin in non-primate mammalian species. Highly purified relaxin was used to (1) determine the primary structure of relaxin, (2) develop specific and sensitive homologous radioimmunoassays that determined the secretory profiles of relaxin during pregnancy, (3) identify vital physiological roles of relaxin during pregnancy in rats and pigs, (4) identify target tissues for relaxin and (5) conduct studies aimed at determining the mechanism of action of relaxin.[3] These studies revealed the remarkable diversity among species that exists with the structure, tissue source, regulation of synthesis and secretion, and physiological effects of relaxin.

Beginning in the mid 1980s and continuing until the present, considerable research has been conducted with human relaxin. It began in 1983 when Hudson and colleagues at the Howard Florey Institute of Experimental Physiology and Medicine in Melbourne used a porcine relaxin cDNA probe to identify a genomic clone encoding biologically active human relaxin.[10] Both synthetic and recombinant human relaxin were then used to develop sensitive immunoassays that determined serum relaxin levels in women during the menstrual cycle and throughout pregnancy.[3] In the mid 1990s Genentech conducted clinical trials to examine the use of recombinant human relaxin as an agent to promote cervical softening at term pregnancy,[11,12] and in the late 1990s Connetics Corp. examined recombinant human relaxin as a therapeutic agent that would prevent progression of systemic sclerosis (scleroderma).[13] Neither clinical trial was successful.

Unlike the 1960s when interest in relaxin declined following the reduction in human clinical research, the frequency of relaxin publications did not decline following unsuccessful clinical trials in the 1990s. This is largely the case because recent discoveries have provided scientifically attractive and clinically promising avenues of investigation that are progressively increasing research on relaxin and related peptides (Fig. 1).

Four major active research areas are described by leading investigators in this book. First, three chapters are devoted to biological actions of endogenous relaxin. Both established and more recently identified putative roles of relaxin on the female reproductive tract during pregnancy are described by Parry and Vodstrcil (Chapter 4). Relaxin also has actions on non-reproductive tissues during pregnancy.[14] Emerging evidence that relaxin promotes vasodilation through a local ligand-receptor system and thereby brings about renal and other peripheral circulation adaptations of pregnancy is described by Jeyabalan and Conrad (Chapter 7). Surprising discoveries in male mice that are deficient in either relaxin or the relaxin receptor provide evidence that relaxin has actions in non-pregnant animals.[15-18] Samuel and coworkers (Chapter

6) summarize their findings that relaxin has physiological roles in modulating extracellular matrix turnover as well as potential antifibrotic therapeutic application in non-reproductive organs such as the heart, lung, and kidney.

Research on relaxin-like peptides is emerging as a major research area with INSL3 getting the most experimental attention (see Fig. 1). INSL3 (Ley-I-L, RLF) was discovered in 1993 at the University of Göttingen by Ibraham Adham and colleagues when they cloned a novel insulin-like peptide from a boar testis cDNA library.[19] The gene, which is expressed in the testicular Leydig cells, is now generally referred to as INSL3. In this book, Agoulnik (Chapter 7) describes INSL3's established vital role in male development and includes findings that indicate that relaxin may also have actions on male reproductive organs.

The identification of relaxin and INSL3 receptors, which occurred about five years ago, has enabled rapid progress in a third major research area. Within a two-year period, relaxin and INSL3 receptors were identified. Hsu and Hsueh and colleagues at Stanford University and Agoulnik and coworkers at Baylor College of Medicine discovered that two orphan leucine-rich guanine nucleotide-binding (G-protein)-coupled receptors designated LGR7 and LGR8 are receptors for relaxin[20] and relaxin-like peptide INSL3,[20-22] respectively. Schwabe and Bullesbach (Chapter 2) describe the discovery and structure/activity relationships of relaxin and INSL3, as well as their receptors LGR7 and LGR8, respectively. Then Ivell and coworkers (Chapter 3) review the current state of understanding of the signaling mechanisms initiated by relaxin and INSL3. Recent studies provide evidence that relaxin and INSL3 may bind to their receptors and activate signaling cascades in a tumor specific context. Klonisch and coworkers (Chapter 8) describe several lines of evidence that relaxin/INSL3 ligand-receptor systems impact on tumor cell metastasis and invasion.

A fourth and increasingly active research area began in 2002 when Bathgate and colleagues at the Howard Florey Institute in Melbourne, Australia discovered a relaxin-like molecule designated relaxin-3[23] (see Fig. 1). More recently, Liu and Lovenberg and colleagues at Johnson and Johnson determined that two different orphan G protein-coupled receptors designated GPCR135 and GPCR142 are putative receptors for relaxin-3 and relaxin-like peptide INSL5, respectively.[24,25] Utilizing available vertebrate and invertebrate genomes Wilkinson and Bathgate (Chapter 1) examined the evolution of the relaxin family and their receptors and concluded that the ancestral relaxin system was relaxin-3 acting through its receptor GPCR135. The function of relaxin-3 remains poorly understood, but progress is being made. Ma and Gundlach (Chapter 9) summarize their recent anatomical and functional studies, providing evidence that relaxin-3 has functions in the brain.

The four major ongoing areas of research described in this book will likely attract investigators for years to come, since many basic scientific questions remain unanswered in each area. Moreover, the clear clinical potential of findings has just begun to be explored. There are other reasons to think that research on relaxin and related peptides has a bright future. The availability of structural databases; new experimental techniques; and high quality hormone preparations, antibodies and other reagents enable rigorous investigation. There is a final extremely important reason for optimism. Nearly all of the investigators who have contributed to this book are early- to mid-career scientists. They bring key ingredients of fresh perspective, enthusiasm and new investigative skills to the relaxin and related peptide field.

I will close with personal comments from a long-standing "relaxinologist". First, when I began conducting research on relaxin in the late 1960s the hormone had not been isolated and it received little experimental attention. Moreover, the impure relaxin preparations and, in some cases, inappropriate or lax experimental approaches used for basic and clinical studies of relaxin during the late 1950s, had created a barrier of skepticism in the minds of some onlooking scientists. One of the satisfactions of the last 30 years has been the widely held acceptance that relaxin has vital reproductive roles in some mammalian species, diverse actions in non-reproductive as well as in reproductive organs and therapeutic potential. In order to retain this desired status in the scientific community, it is important that those of us working in the field make only those claims concerning the activities and importance of relaxin that are well substantiated by rigorous publications. There is a second barrier that relaxinologists must deal with. The diversity of structure, source, secretory profiles and physiological roles of relaxin among species is extraordinary. This among-species variation, which can be confusing, creates a special obligation to present relaxin research and its implications carefully and clearly. There are attractive reasons that are to varying degrees scientific, historical and pragmatic for linking relaxin, INSL3, relaxin-3 and other so-called relaxin-like molecules for scientific meetings, reviews and books such as this one. However, available evidence indicates that no other molecule activates relaxin receptor LGR7 or brings about relaxin's bioactivities. Clarity and understanding will be enhanced if only relaxin bear the name relaxin and other hormones that have evolutionary ties with relaxin be named to reflect their unique biological actions once they are known.

Hopefully sufficient resources and other factors will enable the present young generation of active researchers to have gratifying careers conducting research on relaxin and relaxin-like proteins. Good luck will likely play a role. For my laboratory, the findings in pregnant rats that relaxin promotes nipple growth[3] and that a circadian process tightly regulates parturition[3] occurred serendipitously during the course of experiments conducted for other purposes. Perhaps the running title in a 1958 publication by Kliman and Greep[26] entitled "The enhancement of relaxin-induced growth of the pubic ligament in mice" was intentional. The running title reads "The Enchantment of Relaxin"!

O. David Sherwood , Ph.D.

References

1. Hisaw FL. Experimental relaxation of the pubic ligament of the guinea pig. Proc Soc Exp Biol Med 1926; 23:661-663.
2. Fevold HL, Hisaw FL, Meyer RK. The relaxative hormone of the corpus luteum. Its purification and concentration. J Am Chem Soc 1930; 52:3340-3348.
3. Sherwood OD. Relaxin. In: Knobil E, Neill JD, eds. The Physiology of Reproduction. Vol. 2. New York, NY: Raven Press Ltd., 1994:861-1009.

4. Steinetz BG, Beach VL, Kroc RL et al. Bioassay of relaxin using a reference standard: A simple and reliable method utilizing direct measurement of interpubic ligament formation in mice. Endocrinology 1960; 67:102-115.
5. Sherwood OD, O'Byrne EM. Purification and characterization of porcine relaxin. Arch Biochem Biophys 1974; 160:185-196.
6. Sherwood OD. Purification and characterization of rat relaxin. Endocrinology 1979; 104:886-892.
7. John MJ, Borjesson BW, Walsh JR et al. Limited sequence homology between porcine and rat relaxins: Implications for physiological studies. Endocrinology 1981; 108:726-729.
8. Stewart DR, Nevins B, Hadas E et al. Affinity purification and sequence determination of equine relaxin. Endocrinology 1991; 129:375-383.
9. Stewart DR, Henzel WJ, Vandlen R. Purification and sequence determination of canine relaxin. J Protein Chem 1992; 11:247-253.
10. Hudson P, Haley J, John M et al. Structure of a genomic clone encoding biologically active human relaxin. Nature 1983; 301:628-631.
11. Bell RJ, Permezel M, MacLennan A et al. A randomized, double-blind, placebo-controlled trial of the safety of vaginal recombinant human relaxin for cervical ripening. Obstet Gynecol 1993; 82:328-333.
12. Brennand JE, Calder AA, Leitch CR et al. Recombinant human relaxin as a cervical ripening agent. Br J Obstet Gynaecol 1997; 104:775-780.
13. Erikson M S, Unemori EN. Relaxin clinical trials in systemic sclerosis. In: Tregear GW, Ivell R, Bathgate RA, Wade JD, eds. Relaxin 2000. Dordrecht, The Netherlands: Kluwer Academic Publishers, 2001:373-381.
14. Conrad KP, Novak J. Emerging role of relaxin in renal and cardiovascular function. Am J Physiol Regul Integr Comp Physiol 2004; 287:R250-261.
15. Samuel CS, Zhao C, Bathgate RA et al. Relaxin deficiency in mice is associated with an age-related progression of pulmonary fibrosis. FASEB J 2002; 17:121-123.
16. Du XJ, Samuel CS, Gao XM et al. Increased myocardial collagen and ventricular diastolic dysfunction in relaxin deficient mice: a gender phenotype. Cardiovasc Res 2003; 57:395-404.
17. Samuel CS, Zhao C, Bond CP et al. Relaxin-1-deficient mice develop an age related progression of renal fibrosis. Kidney Int 2004; 65:2054-2064.
18. Hewitson TD, Mookerjee I, Masterson R et al. Endogenous relaxin is a naturally occurring modulator of experimental renal tubulointerstitial fibrosis. Endocrinology 2007; 148:660-669.
19. Adham IM, Burkhardt E, Benahmed M et al. Cloning of a cDNA for a novel insulin-like peptide of the testicular Leydig cells. J Biol Chem 1993; 268:26668-26672.
20. Hsu SY, Nakabayashi K, Nishi S et al. Activation of orphan receptors by the hormone relaxin. Science 2002; 295(5555):671-674.
21. Overbeek PA, Gorlov IP, Sutherland RW et al. A transgenic insertion causing cryptorchidism in mice. Genesis 2001; 30(1):26-35.
22. Kumagai J, Hsu SY, Matsumi H et al. INSL3/Leydig insulin-like peptide activates the LGR8 receptor important in testis descent. J Biol Chem 2002; 277(35):31283-31286.
23. Bathgate RAD, Samuel CS, Burazin TCD et al. Human relaxin gene 3 (H3) and the equivalent mouse relaxin (M3) gene. J Biol Chem 2002; 277:1148-1157.
24. Liu C, Eriste E, Sutton S et al. Identification of relaxin-3/INSL7 as an endogenous ligand for the orphan G-protein-coupled receptor GPCR135. J Biol Chem 2003; 278:50754-50764.
25. Liu C, Chen J, Sutton S et al. Identification of relaxin-3/INSL7 as a ligand for GPCR142. J Biol Chem 2003; 278:50765-50770.
26. Kliman B, Greep RO. The enhancement of relaxin-induced growth of the pubic-ligament in mice. Endocrinology 1958; 63:586-595.

Chapter 1

The Evolution of the Relaxin Peptide Family and Their Receptors

Tracey N. Wilkinson and Ross A.D. Bathgate*

Abstract

The relaxin peptide family in humans consists of relaxin-1, 2 and 3 and the insulin-like peptides (INSL)-3, 4, 5 and 6. The evolution of this family has been controversial; points of contention include the existence of an invertebrate relaxin and the absence of a ruminant relaxin. Over the past four years we have performed a comprehensive analysis of the relaxin peptide family using all available vertebrate and invertebrate genomes. Contrary to previous reports an invertebrate relaxin was not found; sequence similarity searches indicate the family emerged during early vertebrate evolution. Phylogenetic analyses revealed the presence of potential relaxin-3, relaxin and INSL5 homologs in fish; dating their emergence far earlier than previously believed. There are four known relaxin peptide family receptors; the relaxin and INSL3 receptors, the leucine rich repeat containing G protein-coupled receptors (LGR), LGR7 and LGR8 respectively; and the two relaxin-3 receptors, GPCR135 and GPCR142. Database searching identified several invertebrate ancestors of LGR7 and LGR8; the absence of an invertebrate relaxin suggests the presence of an unidentified invertebrate ligand for these receptors. No invertebrate ancestors of GPCR135 or GPCR142 were found. Based on the theory that interacting proteins co-evolve together, phylogenetic analyses of the relaxin peptide family receptors were performed to provide insight into interactions within the relaxin system. Co-evolution between INSL5 and GPCR142, as evidenced by the loss of both genes in the rat and dog and their similar expression profiles, predicted GPCR142 to be the endogeneous INSL5 receptor. This interaction has since been confirmed experimentally. The emergence and presence of multiple GPCR135 homologs in fish reflected similar findings for relaxin-3. It seems likely the ancestral relaxin system was relaxin-3 acting through GPCR135, before LGR7 was "acquired" as a relaxin receptor early in vertebrate development.

Introduction

In the fifty years since the discovery of DNA, great strides have been made in our understanding of the genetic basis of life. Arguably the biggest leap forward was made in 2001 when public and private competing groups jointly published draft sequences of the human genome.[1,2] The genomic age had arrived. Its impact on the study of relaxin was immediate, as a search of the human genome uncovered a novel relaxin gene, relaxin-3.[3] Over the next few years, the pace of genome sequencing increased exponentially and a flood of genomes became and continue to become, available. This incredible wealth of data providing the perfect opportunity to take a comparative genomics approach, using bioinformatic techniques, to the study of the relaxin peptides and their newly discovered receptors.

*Corresponding Author: Ross A.D. Bathgate—Howard Florey Institute, University of Melbourne, Victoria 3010, Australia. Email: bathgate@florey.edu.au

Relaxin and Related Peptides, edited by Alexander I. Agoulnik. ©2007 Landes Bioscience and Springer Science+Business Media.

Table 1. Nomenclature of the relaxin peptides across mammals

Humans and Great Apes	All Other Mammals (e.g., rodents)	Gene Name
Relaxin-1	Relaxin-1*	*RLN1*
Relaxin-2*		*RLN2*
Relaxin-3	Relaxin-3	*RLN3*
INSL3	INSL3	*INSL3*
INSL4	INSL4	*INSL4*
INSL5	INSL5	*INSL5*
INSL6	INSL6	*INSL6*

* Relaxin-2 in humans and great apes, is functionally equivalent to relaxin-1 in all other mammals, both are referred to as relaxin throughout this paper.

The Relaxin Peptides

The relaxin peptides are: relaxin-1, relaxin-2, relaxin-3 and the insulin-like (INSL) peptides, INSL3, INSL4, INSL5 and INSL6. For historical reasons, the peptide designated relaxin-2 in humans and the Great Apes is equivalent to relaxin-1 in all other mammals. To reduce confusion; these will both be simply referred to as relaxin throughout (Table 1). All share the basic structural signature of the insulin peptide with six cysteine residues in conserved positions, which confer two inter-chain and one intra-chain disulfide bonds. Thus, it was postulated that relaxin and insulin derived from a common ancestral gene and were therefore grouped as the insulin superfamily, which also includes the insulin-like growth factors I and II (IGF-1 and -2) (reviewed in [4]). Despite less than 50% predicted sequence similarity across members of the relaxin peptide family, primary structural determinants are retained (Fig. 1). The two chain structure is formed by the removal of the C chain from the pro-hormone to form the A-B heterodimer. This processing has been established for relaxin,[5] relaxin-3[6] and INSL3[7] but has not yet been confirmed for H1 relaxin or INSL4-6. The three relaxin peptides are distinguished by the presence of a relaxin-binding cassette (Arg-X-X-X-Arg-X-X-Ile) in the B chain, which is critical for interaction with the relaxin receptor (LGR7/RXFP1, see below).

Relaxin

In 1983 and 1984 two relaxin genes were discovered in humans[8,9] Named relaxin-1 (*RLN1*) and relaxin-2 (*RLN2*), both of these genes have since been discovered in the Great Apes (i.e., human, chimpanzee, orangutan, baboon and gorilla), with *RLN1* being primate specific.[10,11] *RLN1* and *RLN2* are found adjacent on human chromosome 9 in a cluster of relaxin family genes.[12] Although discovered second, it is the relaxin-2 peptide (H2 relaxin in the human) that is the major circulating form of relaxin in humans and the Great Apes and the functional equivalent to the relaxin-1 found in all other mammals (Table 1).[13] To help address the obvious confusion caused by this historical mistake, both H2 relaxin and its equivalent in other mammals, relaxin-1, will be referred to as relaxin throughout this review. The primate specific relaxin-1 has not been found circulating in the blood and mRNA is expressed in only a few tissues in humans; the decidua, placenta and prostate.[14] A synthetic peptide based on the human *RLN1* sequence has similar activity to H2 relaxin on the human relaxin receptor.[15] The function of H1 relaxin and primate relaxin-1 remains unknown. The biological roles of relaxin will be discussed in detail by other authors in this book.

Relaxin-3

Relaxin-3 was only recently discovered in the Celera human genome database.[3] It was identified as a third relaxin peptide based on the presence of the relaxin-binding cassette and six invariant cysteine residues known to confer the structure specific for insulin/relaxin family members. While

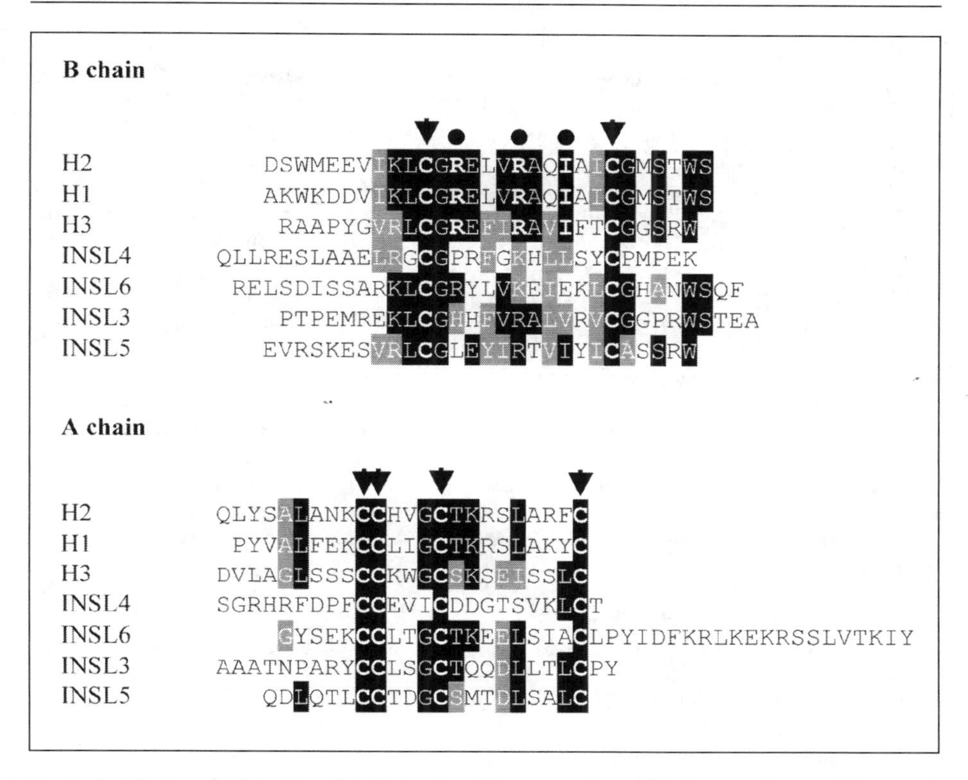

Figure 1. Alignment of the B and A chains for all human relaxin family peptides. B and A chain sequences determined based on known cleavage or predicted cleavage sites. Invariant cysteines are in bold and highlighted with arrows the relaxin-binding cassette residues with circles.

high sequence divergence may be a hallmark of relaxin, a contrastingly high degree of conservation is seen in relaxin-3 sequences.[3]

High relaxin-3 expression is seen in the rodent brain and distribution of rat and mouse relaxin-3 mRNA was determined to be abundant only in the neurons of the nucleus insertus (NI) of the median dorsal tegmental pons.[16,17] These findings indicate that relaxin-3 is likely to be a neuropeptide with roles in the central nervous system quite distinct from the traditional endocrine functions associated with relaxin peptides. The potential roles of relaxin-3 are covered in detail in other parts of this book.

Insulin-Like Peptide 3 (INSL3)

INSL3 was discovered in a differential screening project for porcine testis-specific transcripts and was initially named the Leydig insulin-like peptide.[18] Although located in reasonably close proximity to relaxin-3 on chromosome 19 in the human, *INSL3* is found some distance apart from relaxin-3 on mouse chromosome 8. Sequence similarity between INSL3 and relaxin is significant, particularly in the B chain, although the relaxin binding cassette is displaced in relation to its position in the relaxin B chain. INSL3 has a very low affinity for the relaxin receptor (see below) however it utilizes similar residues for its high affinity binding to its own receptor.[19] The physiological roles of INSL3, especially its actions in mediating testicular descent, will be covered in other chapters of this book.

Insulin-Like Peptide 4 (INSL4)

INSL4 was discovered in 1995 by screening of an early human placenta cDNA library and named the early placenta insulin-like peptide (EPIL).[20] In situ hybridization mapped the *INSL4* gene to human chromosome 9p24 which first revealed the relaxin cluster when it was found in close proximity to *RLN1* and *RLN2*.[21] Present in higher primates only, *INSL4* has been sequenced in the human,[20] identified in the chimpanzee genome and rhesus monkey ESTs,[12] and also been identified using PCR and hybridization analysis in another of the Old World monkeys, the African green monkey.[22] While there is high sequence divergence between relaxin and *INSL4*, *INSL4* sequences across species show relatively low divergence.

The promoter region of *INSL4* was recently found to contain a human endogenous retrovirus (HERV) element, believed to have been inserted 45 million years ago, around the same time as the divergence of *INSL4*.[22] Based on this evidence, the authors postulated that this HERV element mediates the placenta specific expression of *INSL4* and that *INSL4* diverged prior to relaxin-1.[22] *INSL4* mRNA can be translated, in vitro, into both pro-EPIL and EPIL peptides.[23] It is not known whether a processed INSL4 peptide is produced in vivo. Synthetic INSL4 peptides have been produced based on the deduced amino acid sequence and the native structures of other relaxin/insulin family peptides. Structural analysis of these peptides indicates that they are largely unstructured in solution.[24,25] This contrasts with the highly similar and fixed two chain tertiary structures that characterize this family. It is also possible that, like IGFs, INSL4 remains unprocessed in vivo. The receptor for INSL4 is currently unknown.

Insulin-Like Peptide 5 (INSL5)

INSL5 was discovered by two independent groups in EST databases based on sequence similarity with the A and B domains of insulin and relaxin.[26,27] Highest expression of INSL5 in humans is in the colon,[26] while quantitative RT-PCR also detected widespread expression in many human tissues- brain, fetal brain, kidney, prostate, ovary, thymus, bone marrow, placenta, spleen and heart.[28] In the mouse, *Insl5* expression is found in the brain, kidney, testis and thymus.[26,27] In both the human and mouse genomes, the *INSL5* gene is located by itself on chromosomes 1 and 4 respectively.[12]

Within the mouse colon, cells expressing *Insl5* are a discrete population of enteroendocrine-like cells in the crypts.[29] These cells are not found within the upper gastrointestinal tract, instead spanning the colonic epithelial layer.[29] Transgenic mice overexpressing *Insl5* displayed no discernable phenotype,[29] while *INSL5* is a pseudogene in rats and dogs.[28] The function of INSL5 is still unknown; its expression profile suggests a nonreproductive role.

Insulin-Like Peptide 6 (INSL6)

INSL6 was also discovered from searches of EST databases by three independent groups.[27,30,31] It is localized in a cluster of relaxin genes in the human (*RLN1*, *RLN2* and *INSL4* on chromosome 9) and adjacent to relaxin on mouse chromosome 19.[31] Expression of *INSL6* is restricted to the testis in humans,[31] though low levels of expression have been detected in other tissues in the rat (prostate) and mouse (thymus, heart, bowel, kidney and brain).[30,31] While its testis specific expression suggests a possible role in the testis, the function of INSL6 remains uncharacterized. A recent study has demonstrated that the meiotic and postmeiotic germ cells of the testis are the predominant site of *INSL6* expression and that a processed INSL6 peptide can be produced by transfected cells.[32] The INSL6 receptor is unknown and synthetic or recombinant INSL6 peptides show no affinity for any of the relaxin peptide family receptors known to date.[6,33,34]

The Evolution of the Relaxin Peptide Family

Relaxin evolution has confounded researchers for decades. High sequence variability in relaxins across closely related species is a well-known feature of this peptide (Fig. 2). In contrast, startling similarities (almost 100% sequence identity at the amino acid level) have been observed between relaxin peptides sequenced from two very distant species; the pig and two whale species.[35] It

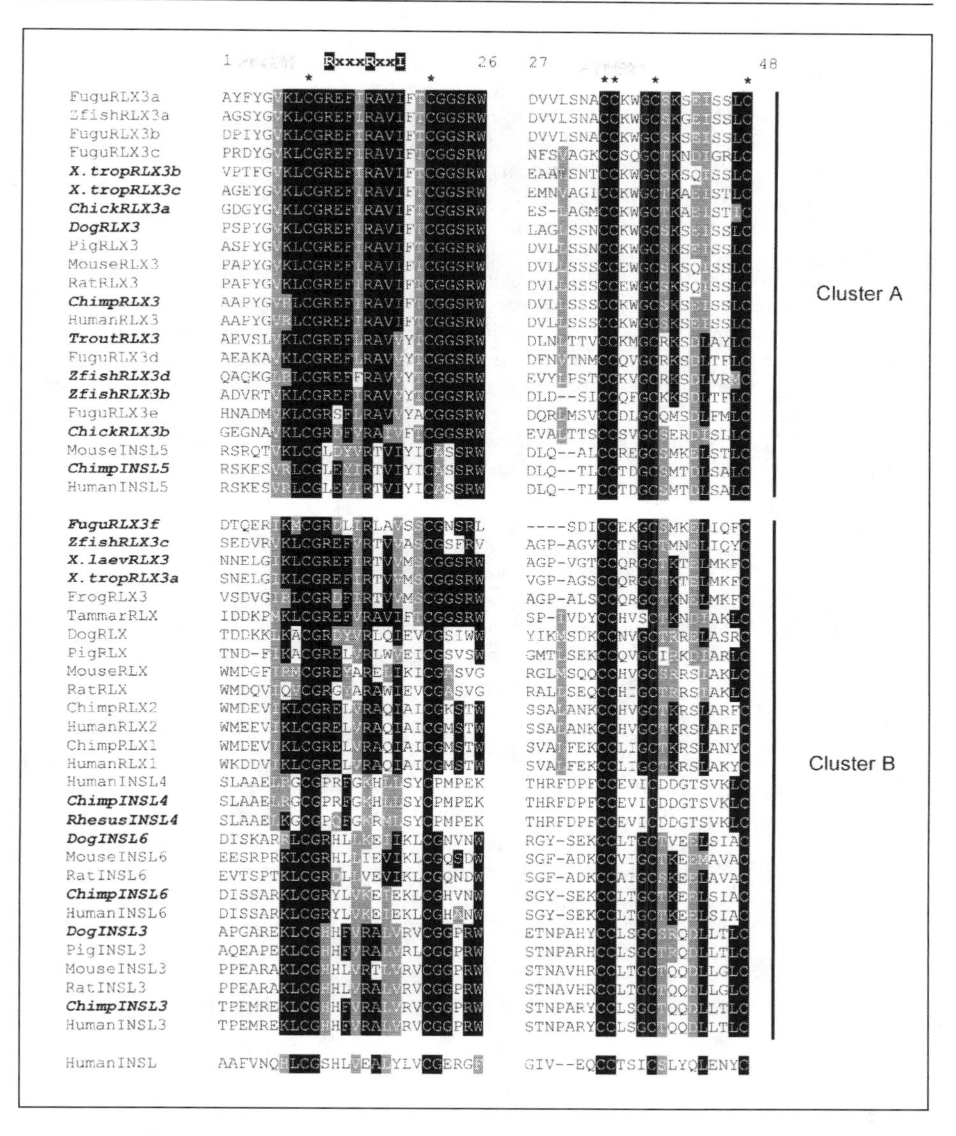

Figure 2. Multiple sequence alignment of the relaxin peptide family. Amino acid sequences of the B and A domains only were aligned using ClustalW, then edited by hand using Seaview to remove gaps. This alignment was then used for all phylogenetic analyses. Newly identified sequences are highlighted in bold and italics. Invariant cysteine residues are indicated by asterisks (*) and the relaxin specific B-chain motif [RxxxRxxI/V] is shown. Sequences are clustered into subfamilies (A and B) based on primary sequence similarity and phylogenetic analysis. X.laev, *Xenopus laevis;* X.trop, *Xenopus tropicalis;* Zfish, Zebrafish.

should be noted that these whale sequences are based on peptide sequencing only and are without supporting nucleotide data. The presence of an invertebrate relaxin has been of speculation since 1983 when relaxin-like activity was first detected in the protozoa, *Tetrahymena pyriformis.*[36] Similar activity was reported in *Herdmania momus* [37] and in *Ciona intestinalis*, where a cDNA sequence almost identical to pig relaxin was found.[38] However, searches of all completed invertebrate genomes (including *C. intestinalis*) failed to identify any relaxin-like sequences, including the published sequence.[12,39] Multiple insulin-like peptides have been found in several invertebrates but as these sequences lack the relaxin-specific motif and show no homology to other relaxin family peptides, they are not considered part of the relaxin peptide family.[12] If an invertebrate relaxin does exist, it does not contain the relaxin-specific motif characterized in vertebrates.

The presence of a functional relaxin in the ruminant lineage has often been postulated. A partial relaxin-like cDNA, with high homology to exon 2 of pig relaxin, has been identified in the ovine.[40] The presence of numerous stop codons in the putative C peptide region and the absence of exon 1 and therefore no B chain, result in a nonfunctional relaxin gene. Searching of the ovine and bovine genome has not identified a relaxin gene although it is possible that the gene still exists in an unsequenced part of the genome. Interestingly, a relaxin-3 ortholog is present in both genomes,[12] and it is possible that expression of this gene may compensate for the loss of relaxin in these species. It has been previously postulated that in fact INSL3 may compensate for the lack of relaxin in ruminants.[41]

The presence of an avian relaxin has also been of speculation. While relaxin-like activity has been reported in the chicken,[42] an avian relaxin peptide or gene has not been identified. While two relaxin-3-like genes were identified in the chicken genome, no avian relaxin gene was found. As no other relaxin-like genes were found, the reported relaxin activity may be due to one of the relaxin-3-like genes.[12]

The phylogeny of the relaxin peptide family indicates relaxin-3 is the ancestral relaxin, appearing prior to the divergence of teleosts.[12,43] The finding of multiple relaxin-3-like genes in the fugu fish and zebrafish suggests multiple lineage-specific duplications of a single relaxin-3-like ancestor have occurred in fish.[43] However, the possibility the other mammal specific relaxin-like peptides emerged earlier before being lost in the teleost can not be excluded.[43] It is more likely that these duplications and the resulting multiple relaxin-3-like genes, are fish specific and due to genome wide duplications hypothesized to have occurred during fish evolution. Phylogenetic analyses show multiple fish homologs of both the mammalian relaxin-3 and *INSL5* genes, meaning that two relaxin-3-like genes existed prior to the genome duplication event proposed to have occurred in the teleost ancestor. The putative fish relaxin homolog was either, present in the teleost ancestor, duplicated and the second copy lost or emerged shortly after or, as a result of, the genome-wide duplication event.

Phylogenetic results suggest that a relaxin homolog exists in frogs as well as fish although not in the chicken.[12] A relaxin gene has been cloned from the edible frog [44] and relaxin peptides have been isolated and the peptide sequence determined from either the ovaries or testes of the little skate (*Raja erinacea*),[45] spiny dogfish (*Squalus acanthias*),[46] Atlantic stingray (*Dasyatis sabina*)[47] and the sand tiger shark (*Odontaspis taurus*).[48] While all these peptides show high similarity with relaxin-3, these sequences are not relaxin-3 homologs (as the B chain of the stingray sequence is lacking the relaxin-specific motif, it is unlikely to be a functional relaxin). Based on the expression of all these genes in reproductive organs such as the testes and ovaries and the failure to find the R. esculenta gene expressed in the brain using Northern blot analysis,[44] these peptides could be among the first relaxin peptides with a reproductive function. Based on the similarity with relaxin-3 observed in these sequences, the ancestral relaxin homolog and its new reproductive function, is likely to have emerged prior to the divergence of teleosts. A complete picture of the relaxin family peptides present in nonmammalian genomes will be invaluable in understanding the evolution of relaxin from neuropeptide to reproductive hormone.

The Relaxin Peptide Family Receptors

Despite extensive study since 1926, of relaxin and later the relaxin peptide family, their receptors remained unknown for almost 80 years. Due to the structural similarity between the relaxin-like peptides and insulin, efforts were concentrated on finding a tyrosine kinase receptor with similarity to the insulin receptor. The recent discovery of four relaxin peptide family receptors provided the opportunity to study the relaxin peptides alongside their receptors.

Leucine Rich Repeat Containing G Protein-Coupled Receptor 7 and 8 (LGR7 and LGR8)

It was not until 2002 that the first relaxin family receptors were identified; relaxin was demonstrated to bind and activate two previously orphan heterotrimeric guanine nucleotide binding protein-coupled receptors (GPCR) LGR7 and LGR8.[49] While relaxin can bind to both LGR7 and LGR8, INSL3 is specific for LGR8. To complicate matters, relaxin-3 is also a high affinity agonist for LGR7.[15] Further work has showed that INSL3 is a higher affinity agonist for LGR8 than relaxin,[50,51] and both mouse and rat relaxin will not activate Lgr8.[15,51] Studies of the *Lgr8* knockout mouse model [52] which has a similar cryptorchid phenotype to that of the *Insl3* knockout models [53] conclusively show LGR8 to be the INSL3 receptor. The identity of LGR7 as the relaxin receptor was confirmed with the *Lgr7* knockout mouse, which has impaired nipple development and parturition defects, similar phenotypes to the relaxin knockout mouse.[54,55] No interactions of other relaxin family peptides with either LGR7 or LGR8 have been detected to date.[25,34,49]

The relaxin and INSL3 receptors are members of the LGR family and are class I (rhodopsin-like) GPCRs. Phylogenetic analysis of the LGR family classified it into three subgroups.[56] Subgroup A contains the glycoprotein hormone receptors: thyroid stimulating hormone receptor (TSHR), follicle stimulating hormone receptor (FSHR) and the luteinizing hormone receptor (LHR), all of which contain 9 LRRs and are the most extensively characterized LGRs to date.[56] Subgroup B contains three orphan receptors, LGR4 and LGR5 with 17 LRRs and LGR6, which despite its classification has only 13 LRRs.[56] Subclass C contains LGR7 and LGR8, which each have 10 LRRs, as well as invertebrate LGRs, LGR3 and LGR4 from Drosophila and a Lymnaea stagnalis (pond snail) LGR that has 12 LRRs.[56] The class C LGRs are further distinguished by the presence of an N-terminal low-density-lipoprotein, type A (LDLa) module in their ectodomains and a unique hinge region between the transmembrane region and the LRRs.[56] LGR7 and LGR8 are the only known mammalian GPCRs to contain an LDLa module. While LDLa modules are generally known as modulators of protein-protein interactions, the LDLa modules in LGR7 and LGR8 are not involved in the primary binding of relaxin/INSL3, but rather are essential for ligand induced cAMP signaling.[57] Such a function has not previously been described for an LDLa module, nor has such a mode of GPCR activation.

GPCR135 and GPCR142

Following the discovery of the relaxin and INSL3 receptors, another two GPCRs specific for relaxin-3 were identified, the somatostatin- and angiotensin- like peptide receptor (SALPR or GPCR135) and GPCR142.[6,33] Similarly to LGR7 and LGR8, GPCR135 was also an orphan receptor, identified as a novel GPCR with high expression in the human brain, primarily in the substantia nigra and pituitary, with low expression levels in peripheral tissues.[58] Relaxin-3 was identified as the ligand of GPCR135 after being purified from a porcine brain extract able to stimulate cells overexpressing GPCR135.[6] In situ hybridization was also able to show neuroanatomical colocalization of GPCR135 with relaxin-3 and binding studies indicated a high affinity interaction.[6] Using GPCR135 as a query to search the human genome, another closely related receptor was discovered, GPCR142 (or GPR100).[33] Further analysis showed GPCR142 to also be a high affinity receptor for relaxin-3.[33] Unlike GPCR135, the expression of GPCR142 is widespread and found in brain, kidney, thymus, testis, placenta, prostate, salivary gland, colon and thyroid.[33] As this does not match with expression of relaxin-3, the presence of another ligand for GPCR142 was postulated. Analysis of the relaxin peptides with their receptors demonstrated the co-evolution of GPCR142 with INSL5.[59] Persuasively, both the rat and dog have nonfunctional INSL5 and GPCR142 genes and further work did indeed show INSL5 to be a high affinity agonist for GPCR142.[28] Combined with their co-evolution and similar expression profiles, INSL5 not relaxin-3, is considered the endogenous ligand for GPCR142. Before the identification of the relaxin-3 and INSL5 ligands for GPCR142, or GPR100, it was first characterised as a high affinity receptor for bradykinin and kallidin.[60] However, it has since been demonstrated that neither of these are able to activate GPCR142 signaling.[61] No interactions with any other insulin or relaxin peptides and GPCR135 or GPCR142 have been detected at this time.[6,33]

Receptor Nomenclature

The international union of pharmacology (IUPHAR) recently suggested that LGR7, LGR8, GPCR135 and GPCR142 be renamed as relaxin family peptide receptors (RXFP) 1, 2, 3 and 4 respectively.[62] This is a reflection of the co-evolution of the peptides and receptors in this family as outlined in Figure 3. It is suggested that irrespective of the nomenclature used in papers by researchers in the field that they refer to both sets of names in the text and keywords of their papers.

The Evolution of the Relaxin Peptide Family Receptors

The complex genomic structures of the LGR genes has made them particularly difficult to identify by sequence similarity searching (Fig. 4). However type C LGRs have been identified in invertebrate genomes[43] and more recently in the mosquito and Ciona genomes (Table 2; unpublished data). Although the sequences contain an LDLa module phylogenetic analysis indicates that these invertebrate sequences are unlikely to be functional orthologs of LGR7 or LGR8.[43,59] This correlates with the absence of relaxin peptides in invertebrates and is supported by preliminary

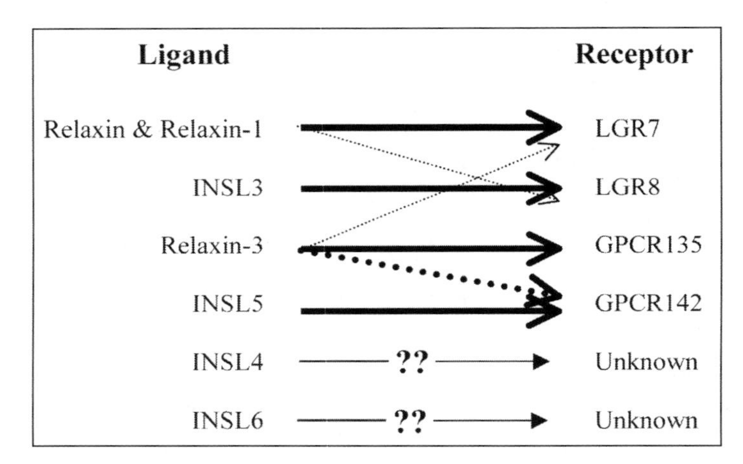

Figure 3. Ligand-receptor pairings of the relaxin peptide family. Endogeneous interactions are shown in solid lines, pharmacological interactions with unknown physiological relevance are shown in dashed lines. The width of the line represents the strength of the interaction.

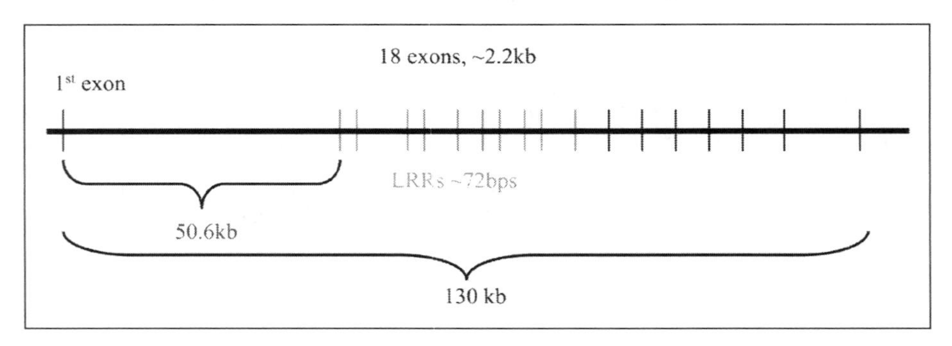

Figure 4. Genomic organization of LGR7 and LGR8. Distances given are based on the human LGR7 sequence, which is representative of all LGR7 and LGR8 sequences.

experimental results that the human relaxin, relaxin-3 nor INSL3 peptides are able to bind or activate Fly (Drosphila melanogaster) LGR4 (unpublished data). In contrast phylogenetic analysis of *LGR7* and *LGR8* sequence identified in fish genomes does demonstrate that these sequences are *LGR7* and *LGR8* orthologs.[59] Importantly, analysis of all available vertebrate genomes has identified *LGR7* and *LGR8* orthologs in all species except the chicken (Table 2).[59] Searches of the chicken genome have not identified a full length *LGR8* gene but this may be due to incomplete sequencing (unpublished data). Interestingly *LGR7* orthologs are found in the sheep and cow genomes despite the absence of a relaxin gene in these species.

Table 2. Results of sequence similarity searches for LGR7 and LGR8

Species	LGR7	LGR8
Human	AF190500, Q9HBX9	AF403384, Q8WXD0
Chimpanzee	Genomic prediction	Genomic prediction
Orangutan	CR860744, CAH92858	?
Rhesus	Genomic prediction	Genomic prediction[1]
Cow	Genomic prediction & ESTs	Genomic prediction, ESTs & cloned sequence[1]
Mouse	AY509975, AAR97515	AF346501, Q91ZZ5
Rat	AY509976, AAR97516	AAW84088, AY906861
Pig	ESTs[1]	ESTs[1]
Dog	Genomic prediction[1]	AY749634, AAU95071
Opossum	Genomic prediction	Genomic prediction
Chicken	ESTs	Genomic prediction[1]
Xenopus	Genomic prediction[1]	Genomic prediction[1]
Zebrafish	Genomic prediction & cloned sequence[1]	Genomic prediction & ESTs[1]
Fugu	Genomic prediction	Genomic prediction[1]
Tetraodon	Genomic prediction	Genomic prediction
Fly	NM_170236, AAF56490[2] LGR3 NM_132635, AAF48237[2] LGR4	
Snail	Z23104, P46023[2]	
Ciona	Genomic prediction[1,2]	
Mosquito	ESTs[1,2]	

Accession numbers are provided for sequences present in GenBank (the nucleotide accession is shown first), supporting evidence for each sequence is detailed. Accession numbers for relevant genomic scaffolds and ESTs used are provided below. ? represents a gene likely missing due to incomplete genome sequencing. [1]incomplete sequence, missing first exon and some internal exons due to gaps in genome scaffolds. [2]likely ancestral invertebrate type C LGR sequences not orthologous to LGR7 or LGR8. **Sequence details:** S = genomic scaffold. **Rhesus:** LGR7 (S55532), LGR8 (S31298, S51770, S120728); **Opossum:** LGR7 (S15231), LGR8 (S15105); **Chimpanzee:** LGR8 (S34718); **Cow:** LGR7 (S3301, S16057, S191 & ESTs: DN273942, DN277031, DN276780), LGR8 (sequenced clone in our laboratory, S2135, S3368 & ESTs: AJ690928, AV592705, AV592706, AW306854, BF040869, CN437953); **Pig:** LGR7 (ESTs: BF191498, CA994861, CA994862), LGR8 (ESTs: CA997681, AW436170, AJ683362); **Dog:** LGR7 (contig 24829); **Chicken:** LGR7 (ESTs: CR387456, BU274919, BU366143, BU366385, CO419618, BU274328, BU284397, BU279098, CD218149, BM425549), LGR8 (contig111.152, 111.153, 111.154); **Zebrafish:** LGR8 (S1292, ESTs: BG304121, BG306216, BF158267), LGR7 (sequenced clone in our laboratory, S403); **Tetraodon:** LGR7 (S14744), LGR8 (S14601); **Xenopus:** LGR8 (S554), LGR7 (S231); **Fugu:** LGR7 (S1619), LGR8 (S4226); **Ciona:** chr03p:3,275,613-3,280,062; **Mosquito:** (ESTs: BM616490, BM614836).

Table 3. Results of sequence similarity searches for GPCR135 and GPCR142

Species	GPCR135	GPCR142
Human	D88437, Q9NSD7	AY394502, Q8TDU9
Chimpanzee	Genomic prediction	Genomic prediction
Rhesus	Genomic prediction	Genomic prediction[1]
Macaca	?	AY633766, AAU93892
Mouse	AY633762 Q8BGE9,	AY633765, Q7TQP4
Rat	AY633764, AAU93890	pseudogene
Dog	Genomic prediction	pseudogene
Cow	Genomic prediction	AY633767, AAU93893
Cat	Genomic prediction	
Pig	?	AY633768, AAU93894
Opossum	Genomic prediction	pseudogene
Chicken	Genomic prediction	–
Xenopus	Genomic predictions (2)	–
Tetraodon	Genomic predictions (5)	–
Fugu	AY288410 (GPCR135a) & Genomic predictions (5)	–
Zebrafish	Genomic predictions (5)	–

Accession numbers are provided for sequences present in GenBank (the nucleotide accession is shown first). ? represents a gene likely missing due to incomplete genome sequencing; - represents the absence of an orthologous sequence; a number in brackets represents the number of multiple GPCR135-like sequences found. The genomic scaffolds that contain the predicted sequences are listed below. [1]incomplete sequence due to gap in genome contig. *Genomic scaffolds containing the predicted sequences:* **Chimpanzee:** GPCR135 (S37626), GPCR142 (S36950); **Rhesus:** GPCR135 (S105485), GPCR142 (S45263); **Dog:** GPCR135 (Contig 17082); **Cow:** GPCR135 (Contig 69219); **Opossum:** GPCR135 (S3019); **Cat:** GPCR135 (AC144371); **Xenopus:** GPCR135a (S6147), GPCR135b (S820); **Zebrafish:** GPCR135a (S1748), GPCR135b (S650), GPCR135c (S615), GPCR135d (S1832), GPCR135e (S2468); **Fugu:** GPCR135b (S7), GPCR135c (S1445), GPCR135d (S287), GPCR135e (S767); **Tetraodon:** GPCR135a (S14533), GPCR135b (S14581), GPCR135c (S8807), GPCR135d (S14640), GPCR135e (S14577).

All known *GPCR135* and *GPCR142* genes are intronless and this was found to hold true for all the orthologs identified. No *GPCR135* or *GPCR142* orthologs are present in invertebrate genomes. In contrast multiple *GPCR135*-like sequences have been identified in all fish genomes studied, as well as the Xenopus (Table 3).[59,61] These results reflect the multiple relaxin-3-like sequences also seen in these genomes. These duplications are likely the result of the genome wide duplication events thought to have occurred in the fish lineage. Similar to the relaxin peptide family, *GPCR135* appears to have emerged early in vertebrate evolution, likely prior to the divergence of teleosts. The co-evolution of relaxin-3 with *GPCR135* supports the current belief that *GPCR135* is the ancestral endogenous relaxin-3 receptor.[63] Despite the ability of relaxin-3 to also bind with high affinity to GPCR142,[33] *GPCR142* first appears in eutherians and therefore is unlikely to be the endogenous receptor for relaxin-3, which emerged in fish.

Co-Evolution of the Relaxin Peptide Family and Their Receptors

The relaxin peptide family provides an interesting example of receptor-ligand co-evolution. The phylogenies of both relaxin peptide family receptor types and that of the relaxin peptide family have been combined in a simplified representative tree and the known interactions highlighted in Figure 5. While the relaxin-3/INSL5 branch of the peptide tree co-evolves closely with the GPCR135 and GPCR142 receptor phylogeny, the relaxin/INSL3/INSL4/INSL6 branch of

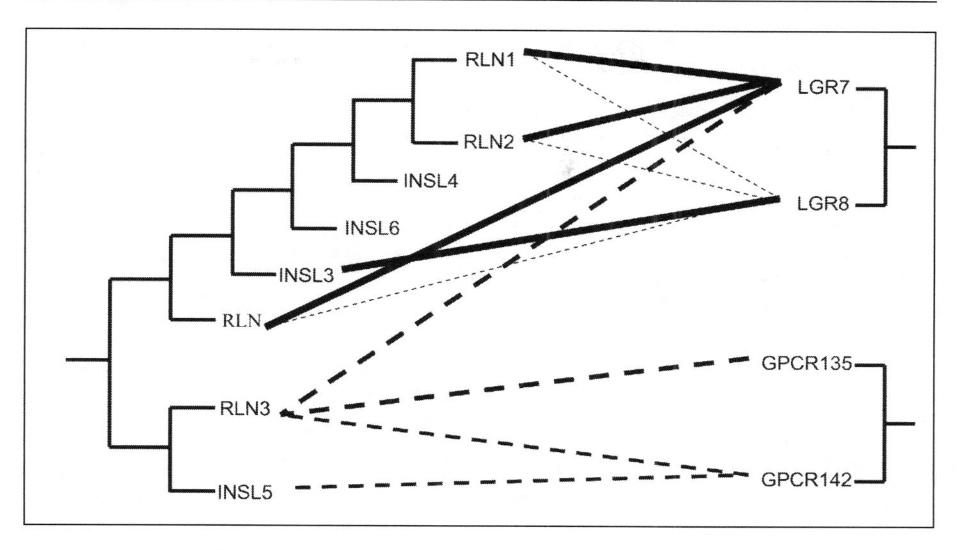

Figure 5. Co-evolution of the relaxin peptides and their receptors. All known interactions are shown, with the predicted endogenous interactions known to be physiologically functional highlighted in bold. High affinity interactions with only pharmacological data are shown in bold dotted lines and lower affinity interactions of unknown functional relevance are shown in thin dotted lines. Simplified representative phylogenies are shown. RLN: Ancestral relaxin which acquired LGR7.

the peptide phylogeny is contrastingly complex, which does not correspond with the LGR7 and LGR8 phylogeny.

The co-evolution of relaxin-3 with GPCR135 is demonstrated by the multiple duplications of both genes during the fish lineage and their correlated emergence, both prior to fish. Combined with pharmacological characterization of their high affinity interaction, their overlapping expression profiles, these results strongly indicate that GPCR135 is the endogenous relaxin-3 receptor. Based on this analysis, which assumes all relaxin-3 receptors are physiologically relevant, relaxin-3 acquired GPCR142 as a receptor following its divergence from GPCR135 and somehow also acquired the unrelated LGR7 as a receptor. Following this event, the relaxin peptides that diverge from relaxin-3 also act through LGR7 and then LGR8, before losing the ability to interact with GPCR135 and GPCR142. Experimental verification of the ability of relaxin peptides to activate the nonmammalian LGR7 and LGR8 sequences is required to place the timing of this event.

References

1. Lander ES, Linton LM, Birren B et al. Initial sequencing and analysis of the human genome. Nature 2001; 409(6822):860-921.
2. Venter JC, Adams MD, Myers EW et al. The sequence of the human genome. Science 2001; 291(5507):1304-1351.
3. Bathgate RA, Samuel CS, Burazin TC et al. Human relaxin gene 3 (H3) and the equivalent mouse relaxin (M3) gene. Novel members of the relaxin peptide family. J Biol Chem 2002; 277(2):1148-1157.
4. Bathgate RAD, Hsueh AJ, Sherwood OD. Physiology and Molecular Biology of the Relaxin Peptide Family. In: Neill JD, ed. Physiology of Reproduction. 3rd ed. San Diego: Elsevier 2006:679-770.
5. Sherwood CD, O'Byrne EM. Purification and characterization of porcine relaxin. Arch Biochem Biophys 1974; 160(1):185-196.
6. Liu C, Eriste E, Sutton S et al. Identification of relaxin-3/INSL7 as an endogenous ligand for the orphan G-protein-coupled receptor GPCR135. J Biol Chem 2003; 278(50):50754-50764.
7. Bullesbach EE, Schwabe C. The primary structure and the disulfide links of the bovine relaxin-like factor (RLF). Biochemistry 2002; 41(1):274-281.

8. Hudson P, Haley J, John M et al. Structure of a genomic clone encoding biologically active human relaxin. Nature 1983; 301(5901):628-631.
9. Hudson P, John M, Crawford R et al. Relaxin gene expression in human ovaries and the predicted structure of a human preprorelaxin by analysis of cDNA clones. EMBO J 1984; 3(10):2333-2339.
10. Evans BA, Fu P, Tregear GW. Characterization of two relaxin genes in the chimpanzee. J Endocrinol 1994; 140(3):385-392.
11. Evans BA, Fu P, Tregear GW. Characterization of primate relaxin genes. Endocr J 1994; 2:81-86.
12. Wilkinson TN, Speed TP, Tregear GW, Bathgate RAD. Evolution of the Relaxin-like peptide family. BMC Evol Biol 2005; 5:14.
13. Winslow JW, Shih A, Bourell JH et al. Human seminal relaxin is a product of the same gene as human luteal relaxin. Endocrinology 1992; 130(5):2660-2668.
14. Hansell DJ, Bryant-Greenwood GD, Greenwood FC. Expression of the human relaxin H1 gene in the decidua, trophoblast and prostate. J Clin Endocrinol Metab 1991; 72(4):899-904.
15. Bathgate RAD, Lin F, Hanson NF et al. Relaxin-3: Improved synthesis strategy and demonstration of its high affinity interaction with the relaxin receptor LGR7 both in vitro and in vivo. Biochemistry 2006; 45:1043-1053.
16. Burazin TC, Bathgate RA, Macris M et al. Restricted, but abundant, expression of the novel rat gene-3 (R3) relaxin in the dorsal tegmental region of brain. J Neurochem 2002; 82(6):1553-1557.
17. Tanaka M, Iijima N, Miyamoto Y et al. Neurons expressing relaxin 3/INSL 7 in the nucleus incertus respond to stress. Eur J Neurosci 2005; 21(6):1659-1670.
18. Adham IM, Burkhardt E, Benahmed M et al. Cloning of a cDNA for a novel insulin-like peptide of the testicular Leydig cells. J Biol Chem 1993; 268(35):26668-26672.
19. Rosengren KJ, Zhang S, Lin F et al. Solution structure and characterization of the receptor binding surface of insulin-like peptide 3. J Biol Chem 2006; 38:28287-28295.
20. Chassin D, Laurent A, Janneau JL et al. Cloning of a new member of the insulin gene superfamily (INSL4) expressed in human placenta. Genomics 1995; 29(2):465-470.
21. Veitia R, Laurent A, Quintana-Murci L et al. The INSL4 gene maps close to WI-5527 at 9p24.1—>p23.3 clustered with two relaxin genes and outside the critical region for the monosomy 9p syndrome. Cytogenet Cell Genet 1998; 81(3-4):275-277.
22. Bieche I, Laurent A, Laurendeau I et al. Placenta-specific INSL4 expression is mediated by a human endogenous retrovirus element. Biol Reprod 2003; 68(4):1422-1429.
23. Bellet D, Lavaissiere L, Mock P et al. Identification of pro-EPIL and EPIL peptides translated from insulin-like 4 (INSL4) mRNA in human placenta. J Clin Endocrinol Metab 1997; 82(9):3169-3172.
24. Bullesbach EE, Schwabe C. Synthesis and conformational analysis of the insulin-like 4 gene product. J Pept Res 2001; 57(1):77-83.
25. Lin F, Otvos Jr. L, Kumagai J et al. Synthetic human insulin 4 does not activate the G-protein-coupled receptors LGR7 or LGR8. J Pept Sci 2004; 10(5):257-264.
26. Conklin D, Lofton-Day CE, Haldeman BA et al. Identification of INSL5, a new member of the insulin superfamily. Genomics 1999; 60(1):50-56.
27. Hsu SY. Cloning of two novel mammalian paralogs of relaxin/insulin family proteins and their expression in testis and kidney. Mol Endocrinol 1999; 13(12):2163-2174.
28. Liu C, Kuei C, Sutton S et al. INSL5 is a high affinity specific agonist for GPCR142 (GPR100). J Biol Chem 2005; 280(1):292-300.
29. Jaspers S, Lok S, Lofton-Day CE et al. The genomics of INSL5. In: Tregear GW, Ivell R, Bathgate RA, Wade JD, eds. Relaxin 2000: Proceedings of the third international conference on relaxin and related peptides. Dordrecht: Kluwer, 2001.
30. Kasik J, Muglia L, Stephan DA et al. Identification, chromosomal mapping and partial characterization of mouse Insl6: a new member of the insulin family. Endocrinology 2000; 141(1):458-461.
31. Lok S, Johnston DS, Conklin D et al. Identification of INSL6, a new member of the insulin family that is expressed in the testis of the human and rat. Biol Reprod 2000; 62(6):1593-1599.
32. Lu C, Walker WH, Sun J et al. Insulin-like peptide 6 (Insl6): Characterization of secretory status and posttranslational modifications. Endocrinology 2006.
33. Liu C, Chen J, Sutton S et al. Identification of relaxin-3/INSL7 as a ligand for GPCR142. J Biol Chem 2003; 278(50):50765-50770.
34. Bogatcheva NV, Truong A, Feng S et al. GREAT/LGR8 is the only receptor for insulin-like 3 peptide. Mol Endocrinol 2003; 17(12):2639-2646.
35. Schwabe C, Bullesbach EE, Heyn H et al.Cetacean relaxin. Isolation and sequence of relaxins from Balaenoptera acutorostrata and Balaenoptera edeni. J Biol Chem 1989; 264(2):940-943.
36. Schwabe C, LeRoith D, Thompson RP et al. Relaxin extracted from protozoa (Tetrahymena pyriformis). Molecular and immunologic properties. J Biol Chem 1983; 258(5):2778-2781.

37. Georges D, Viguier-Martinez MC, Poirier JC. Relaxin-like peptide in ascidians. II. Bioassay and immunolocalization with anti-porcine relaxin in three species. Gen Comp Endocrinol 1990; 79(3):429-438.
38. Georges D, Schwabe C. Porcine relaxin, a 500 million-year-old hormone? the tunicate Ciona intestinalis has porcine relaxin. FASEB J 1999; 13(10):1269-1275.
39. Hafner M, Korthof G. Does a "500 million-year-old hormone" disprove Darwin? FASEB J 2006; 20(9):1290-1292.
40. Roche PJ, Crawford RJ, Tregear GW. A single-copy relaxin-like gene sequence is present in sheep. Mol Cell Endocrinol 1993; 91(1-2):21-28.
41. Bathgate R, Balvers M, Hunt N et al. Relaxin-like factor gene is highly expressed in the bovine ovary of the cycle and pregnancy: sequence and messenger ribonucleic acid analysis. Biol Reprod 1996; 55(6):1452-1457.
42. Brackett KH, Fields PA, Dubois W et al. Relaxin: an ovarian hormone in an avian species (Gallus domesticus). Gen Comp Endocrinol 1997; 105(2):155-163.
43. Hsu SY. New insights into the evolution of the relaxin-LGR signaling system. Trends Endocrinol Metab 2003; 14(7):303-309.
44. de Rienzo G, Aniello F, Branno M et al. Isolation and characterization of a novel member of the relaxin/insulin family from the testis of the frog Rana esculenta. Endocrinology 2001; 142(7):3231-3238.
45. Bullesbach EE, Schwabe C, Callard IP. Relaxin from an oviparous species, the skate (Raja erinacea). Biochem Biophys Res Commun 1987; 143(1):273-280.
46. Bullesbach EE, Gowan LK, Schwabe C et al. Isolation, purification and the sequence of relaxin from spiny dogfish (Squalus acanthias). Eur J Biochem 1986; 161(2):335-341.
47. Bullesbach EE, Schwabe C, Lacy ER. Identification of a glycosylated relaxin-like molecule from the male Atlantic stingray, Dasyatis sabina. Biochemistry 1997; 36(35):10735-10741.
48. Reinig JW, Daniel LN, Schwabe C et al. Isolation and characterization of relaxin from the sand tiger shark (Odontaspis taurus). Endocrinology 1981; 109(2):537-543.
49. Hsu SY, Nakabayashi K, Nishi S et al. Activation of orphan receptors by the hormone relaxin. Science 2002; 295(5555):671-674.
50. Kumagai J, Hsu SY, Matsumi H et al. INSL3/Leydig insulin-like peptide activates the LGR8 receptor important in testis descent. J Biol Chem 2002; 277(35):31283-31286.
51. Scott DJ, Fu P, Shen PJ et al. Characterization of the Rat INSL3 Receptor. Ann NY Acad Sci 2005; 1041:13-16.
52. Overbeek PA, Gorlov IP, Sutherland RW et al. A transgenic insertion causing cryptorchidism in mice. Genesis 2001; 30(1):26-35.
53. Nef S, Parada LF. Cryptorchidism in mice mutant for Insl3. Nat Genet 1999; 22(3):295-299.
54. Krajnc-Franken MA, van Disseldorp AJ, Koenders JE et al. Impaired nipple development and parturition in LGR7. Mol Cell Biol 2004; 24(2):687-696.
55. Kamat AA, Feng S, Bogatcheva NV et al. Genetic targeting of relaxin and Insl3 receptors in mice. Endocrinology 2004.
56. Hsu SY, Kudo M, Chen T et al. The three subfamilies of leucine-rich repeat-containing G protein-coupled receptors (LGR): identification of LGR6 and LGR7 and the signaling mechanism for LGR7. Mol Endocrinol 2000; 14(8):1257-1271.
57. Scott DJ, Layfield S, Yan Y et al. Characterization of novel splice variants of LGR7 and LGR8 reveals that receptor signaling is mediated by their unique LDLa modules. J Biol Chem 2006; 281:34942-34954.
58. Matsumoto M, Kamohara M, Sugimoto T et al. The novel G-protein coupled receptor SALPR shares sequence similarity with somatostatin and angiotensin receptors. Gene 2000; 248(1-2):183-189.
59. Wilkinson TN, Speed TP, Tregear GW et al. Coevolution of the relaxin-like peptides and their receptors. Ann N Y Acad Sci 2005; 1041:534-539.
60. Boels K, Schaller HC. Identification and characterisation of GPR100 as a novel human G-protein-coupled bradykinin receptor. Br J Pharmacol 2003; 140(5):932-938.
61. Chen J, Kuei C, Sutton SW et al. Pharmacological characterization of relaxin-3/INSL7 receptors GPCR135 and GPCR142 from different mammalian species. J Pharmacol Exp Ther 2005; 312(1):83-95.
62. Bathgate RA, Ivell R, Sanborn BM et al. International Union of Pharmacology: Recommendations for the nomenclature of receptors for relaxin family peptides. Pharmacol Rev 2006; 58:7-31.
63. Liu C, Bonaventure P, Sutton SW et al. Recent progress in relaxin-3-related research. Ann NY Acad Sci 2005; 1041:47-60.

Chapter 2

Relaxin, the Relaxin-Like Factor and Their Receptors

Christian Schwabe* and Erika E. Büllesbach

Abstract

In 1926 Frederick Hisaw discovered a blood-borne factor in pregnant guinea pigs that would cause relaxation of the pubic symphysis in virgin females of the species.[1] The relaxin-like factor gene (RLF), also known as insulin-like 3 (INSL3), was recovered from a library of testicular cDNA.[2] The function of RLF as the mediator of testicular positioning in mice was discovered by gene deletion experiments.[3,4] The report that deletion of a G-protein-coupled receptor in a mouse mutant caused cryptorchidism[5] and that relaxin and RLF and their receptors[6,7] were structurally and functionally similar may well have inspired Drs. Hsueh and Sherwood to put LGR7 and relaxin together and thus, after many agonizing years of uncertainty, the relaxin receptor had yielded its identity.[8] LGR8 was recognized as the human version of the RLF receptor and together LGR7 and LGR8, with their respective ligands, opened to detailed investigation the large and important field of G-protein activated leucine-rich repeat receptors. In the process RLF and LGR8 have yielded some general information that might contribute to our knowledge of receptor/ligand interaction, in particular the enigmatic signal initiation process.

Relaxin

Following its discovery relaxin was carried with care and something that can only be called affection through the few ups and many downs of its course as a hormone. Most of the early work was done with the readily isolated porcine relaxin (~1 gram per kg of late pregnant ovaries[9,10]) until recombinant human relaxin[11] and fully synthetic relaxin from several species became available.[12,13] Human relaxin was used to obtain the first and only X-ray structure of the molecule and that pushed research ahead into the structure/function aspects of the hormone.[14]

Invaluable as a crystal structure is for interaction studies, the ligand and the LGR7 receptor had to be brought together by the relatively old fashioned methods of deduction and inference and in the final analysis by a little audacity. Clearly, if so many different species-specific relaxins reacted with the same (mouse) receptor, the receptor-binding sites had to show a modicum of similarity.

The variability of the primary structures was as baffling as was the ability of most relaxins to stimulate receptors in animals of which the ligand sequences varied by more than 50%. In fact, relaxin from the sandtiger shark varied by ~ 70% from the porcine relaxin and from both mouse relaxins but was still able to stimulate symphysis pubis growth in the mouse.[15] An ancient ovoviviparous chondrichtian produced a hormone that worked in a relatively modern mammal! The receptor-binding site had to be similar between all these molecules. But the few similar side chains included unlikely binding site residues such as arginines B13,17 that are in all relaxins in the same relative positions.[16]

Another amino acid present in almost all relaxins was a relative nonentity, a glycine in position A14. This place is taken by isoleucine in insulin and an overlay of the structures shows that the A-chain loop of relaxin lays flat against the core while the side-chain of isoleucine forces the insulin loop to stand up. It is still not known how the Ile-induced loop position inactivates relaxin

*Corresponding Author: Christian Schwabe—Department of Biochemistry and Molecular Biology, Medical University of South Carolina, 173 Ashley Avenue, PO Box 250509, Charleston SC 29425, U.S.A. Email: schwabec@musc.edu

Relaxin and Related Peptides, edited by Alexander I. Agoulnik.

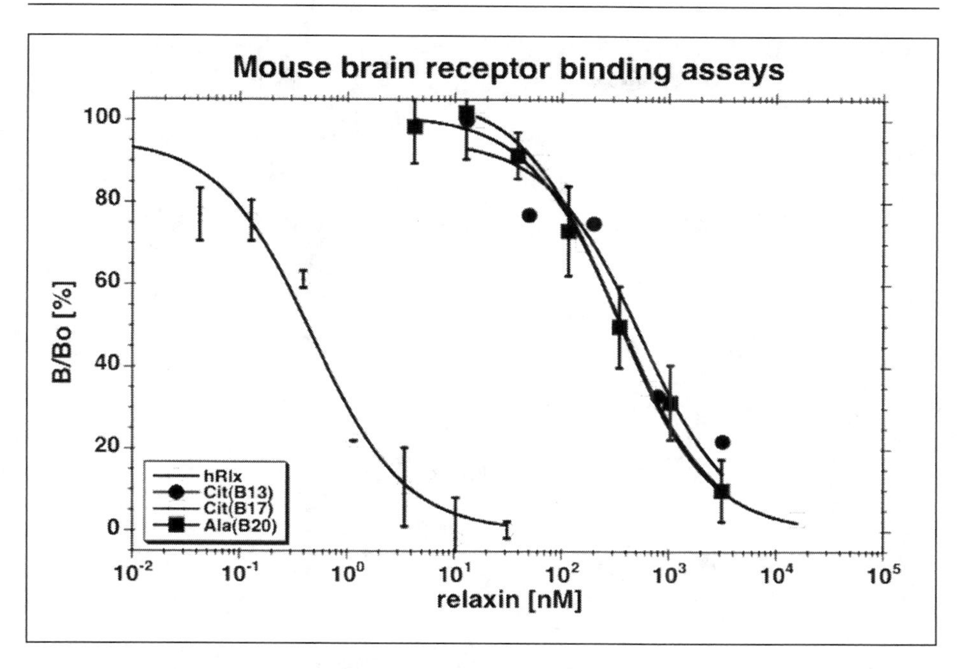

Figure 1. Mouse brain receptor binding of human relaxin and human relaxin derivatives modified in the active site residues B13, B17 or B20.

nor is it known how glycine in the same position would inactivate insulin.[17] As concerns relaxin it is possible that the loop makes contact with the cysteine-rich region of LGR7. Here again relaxin deceived the X-ray crystallographers who maintained that position A14 could accommodate any amino acid without disturbing the molecule because the side-chain would project into the water. Fact of the matter is that the Ile A14 structure forms readily, shows the correct though somewhat blue-shifted CD spectrum and is inactive! There is still no explanation for this observation: biology requires refinements that do not stop at the primary structure level.

The search for the receptor-binding site that is now well publicized provided one of the many surprises that relaxin lavished on the puzzled investigators. When the dust settled all relaxins had arginines in position B13, B17 and a hydrophobic amino acid (Ile or Val) in position B20 (Fig. 1). This is the binding-cassette that became the hallmark of relaxin.[18] It projects from the B-chain helix in a linear arrangement of two arginine 'fingers', followed by isoleucine offset by about 30° like a prehensile thumb (Fig. 2).

This crisp picture of at least the major receptor-binding elements of relaxin encouraged a search for the complementary sites on the receptor of which nothing was known safe the general arrangement as a leucine-rich repeat ecto-domain. Known was the crystal structure of the intracellular porcine RNAse inhibitor, the archetypal leucine-rich repeat protein, in the absence and presence of its substrate.[19,20] With some cutting and splicing the sequence of the leucine-rich repeat region of LGR 7 was accommodated with asparagines at the turn between the pleated sheet of the concave interior lining and a helix in the exterior portion of each repeat. The search for receptor-binding residues was conducted considering the nominally concave binding pocket lined with short parallel pleated sheets transferred to a flat printout of 9 repeats extracted from the LGR 7 sequence.

Targeted were negative charges separated by the translational distance (5.4Å) of an α-helix. The gene of LGR7, kindly provided by Dr. Hsueh, was used in our laboratory for site-directed mutagenesis and the first mutant made on the basis of this crude model was on repeat 8, E277Q and that essentially eliminated receptor-binding.

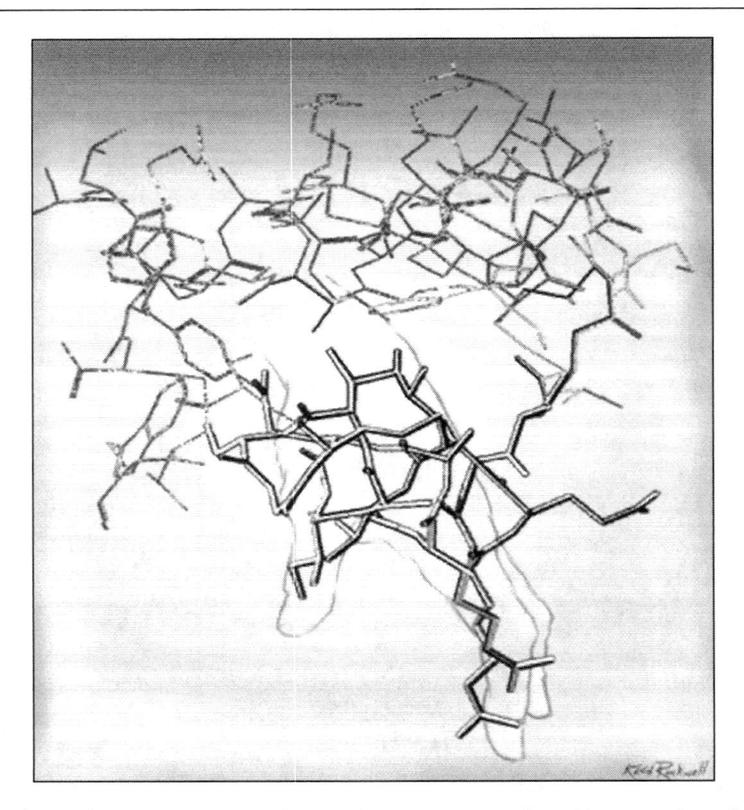

Figure 2. Display of the active site residues in the "gripping mode" of human relaxin[18] drawn on the X-ray coordinates (6rlx.pdb).[14]

It seemed that a simple salt link would be insufficient for binding but the paper model showed another acidic residue located on the same repeat (Asp279), pointing up and separated from E277 by a leucine residue which pointed down below the plane of the receptor surface. Exchange of this aspartic acid for asparagine also eliminated recognition of relaxin by LGR7. It seemed clear that one arginine, later assigned to be Arg(B13), of relaxin was chelated by two acidic residues.

One helix turn, 5.4 Å away toward the N terminus on repeat 6 there were two negative charges Glu231 and Asp233 that, according to our model, made good candidates for contacts with the second arginine. The leucine 232 again points down and out of the way and was of little concern. The result of site-directed mutagenesis showed that the acidic residues Glu231 and Asp233 indeed chelated the second arginine, later recognized as Arg B17.

The isoleucine B20 (the thumb) reached into a hydrophobic region formed by tryptophan 180, isoleucine 182 and leucine 204 on repeats 4 and 5. Exchange of any of these residues again curtailed binding severely. The hydrophobic interaction fixed the direction of relaxin as it crosses the concave surface of LGR7 such that the C-terminal end of relaxin points toward the N-terminal end of the receptor. Residues located between the contacts on LGR7 were replaced without causing interference with receptor binding (Fig. 3).

The relaxin interaction site is complex enough, considering only the precision of placement of the negatively charged groups. Further experiments suggested that within each chelating pair the aspartic acid and the glutamic acid cannot be exchanged without loss of ligand-binding.

It is likely that the N-terminal cysteine-rich region (1-91) supports the relaxin/LGR7 interaction. The cysteine-rich domain on LGR7 comprises two segments. Sequence 1-40 contains six

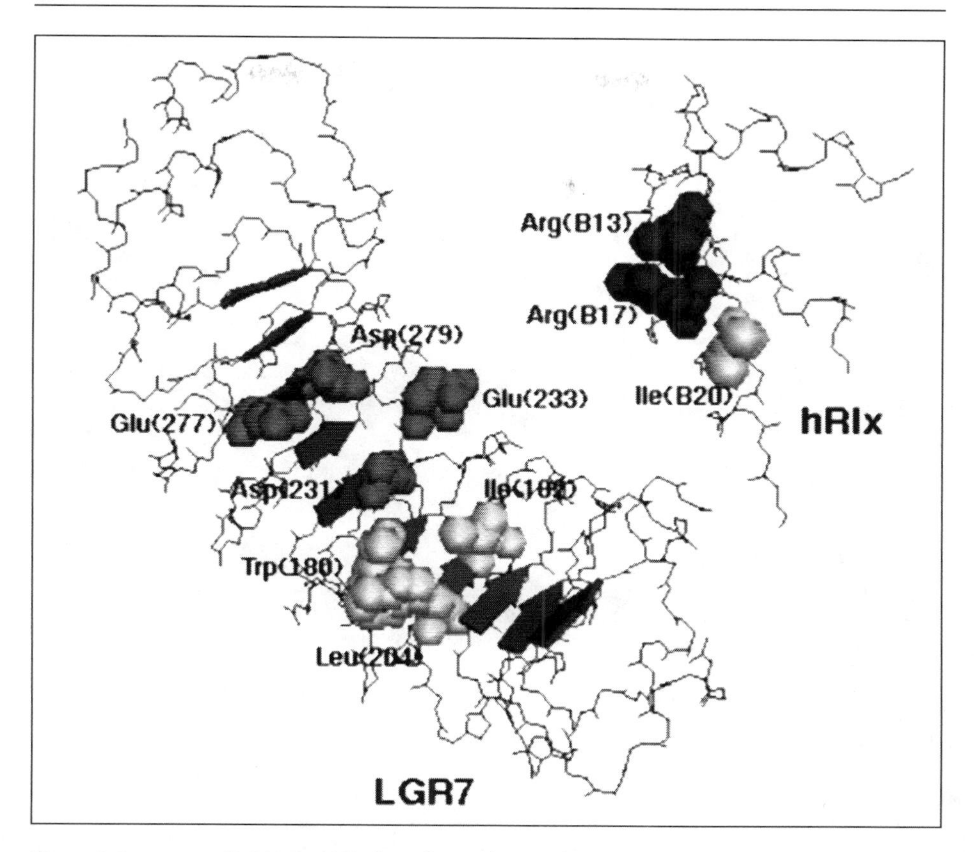

Figure 3. Structure of LGR7 modeled on ribonuclease inhibitor (1dfj.pdb). The hormone contact sites are highlighted in dark grey (acidic residues) and light grey (hydrophobic residues) on the inner circumference. Human relaxin structure (6rlx.pdb)[14] showing the receptor binding site (arginine residues B13, B17 dark grey) and (hydrophobic residue Ile B20, light grey).[22]

cysteines and is homologous to the low-density lipoprotein type A sequence (LDLa) whereas segment 73-91 contains four cysteines. While deletion of the first segment (1-41) has no impact on relaxin-binding[21] the deletion of the complete cysteine-rich region eliminates ligand recognition.[22] It has been suggested that the N-terminal LDLa sequence initiates signal transduction. Details of that action are not known but the coexpression of full length LGR7 shows inhibitory properties.[21]

The failure of clinical trials of recombinant human relaxin as an aid to parturition[23] seemed like the last curtain call for a hormone that appeared to have many functions. Sure enough, other leads were soon reported and it became apparent that relaxin might be a hormone that increases peripheral blood flow that protects the ischemic heart[24] and has a long-term anti-fibrosis action in mice.[25] These are wonderful prospects, which will remain just that until hardcore, long-term clinical trials have been done.

Relaxin purportedly does for humans what no other agent will do. These authors are neither aware of another drug that will increase peripheral blood flow as documented as a byproduct of the scleroderma trial[26] nor of any agent that suppresses age related fibrosis.[27] Heart effects are more difficult to evaluate but all suggest that relaxin might do best as a prophylactic agent.

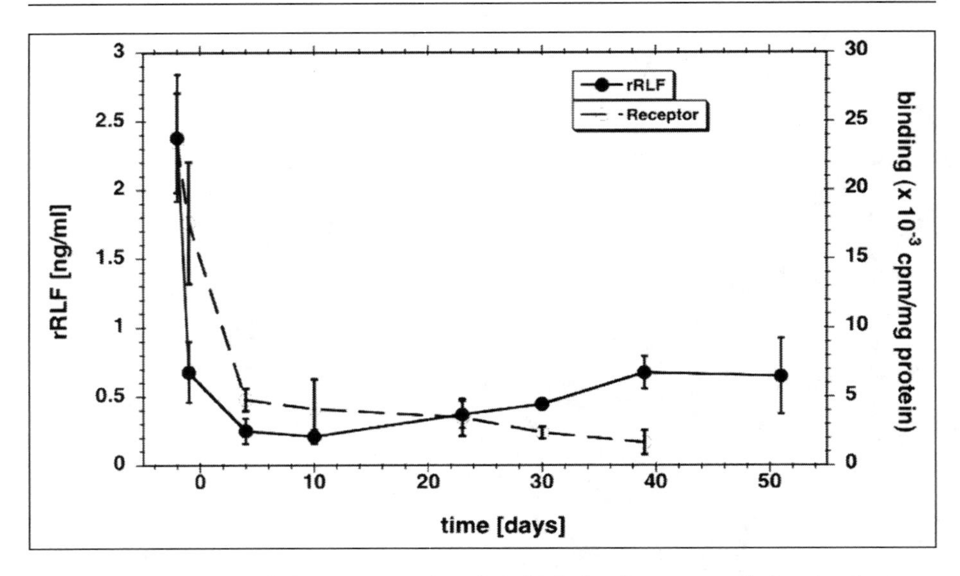

Figure 4. Mean (±SEM) of RLF-concentrations in 100 µl of male rat serum during development (filled circles). Rabbit anti-mouse antiserum 9956 was used in combination with ^{125}I-human RLF as radioligand. Each data-point was collected in duplicate and three independent assays were averaged. The data were compared with a mouse RLF standard curve. RLF-receptor levels on crude membrane preparations of rat gubernacula during development (open circles). The counts per mg of membrane protein were determined in duplicates in three independent experiments. The data were pooled and the mean (±SEM) is displayed.

RLF

Within a few years of its discovery in a porcine testicular cDNA library[2] knockout experiments in mice revealed that the new gene called insulin-like 3 (*Insl3*) was required for normal testicular descent.[3,4] This observation was a breakthrough for the long recognized problem of gonadal retention in human infants which affects about 3% of all newborns but self-corrects to about 1.5% within one year post partum. Loss of fertility can be prevented by surgical intervention but these patients, for unknown causes, retain a statistically significant risk of testicular cancer.[28,29]

The protein was chemically synthesized, characterized and its relaxin-like properties noted; hence the name RLF.[6] Intra-peritoneal injection of synthetic mouse RLF into pregnant *Insl3*$^{-/-}$ mice, paired with heterozygous males did not rescue the cryptorchid phenotype (unpublished). Transgenic female mice expressing *Insl3* in the pancreas actually showed descended ovaries but no rescue for cryptorchid males.[30] Koch's postulate, which, in a different form, is the central paradigm of endocrinology, remained unfulfilled, yet, there clearly had to be a connection between RLF and gonadal retention.

This was a first rate puzzle. Could it be that the protein deduced from the cDNA sequence was not the active form, was RLF processed differently from other hormones, was the pro-form active and the two-chain molecule inhibitory, would RLF have to act as a paracrine or autocrine factor and was therefore not transported to the target tissue or was it just a matter of timing?

Synthetic mouse RLF was used to generate anti-mouse-RLF serum and to determine the level of RLF in male rats pre- and post partum. Prior to parturition (day-2) RLF levels are highest in the male (2.4 ng/ml)[31] and low in the female, confirming reported transcription levels.[32] A sharp drop of RLF occurs at parturition when male and female pups show the same low concentration of RLF in the serum. In the male the RLF level increases slightly at day 10 and reaches a plateau

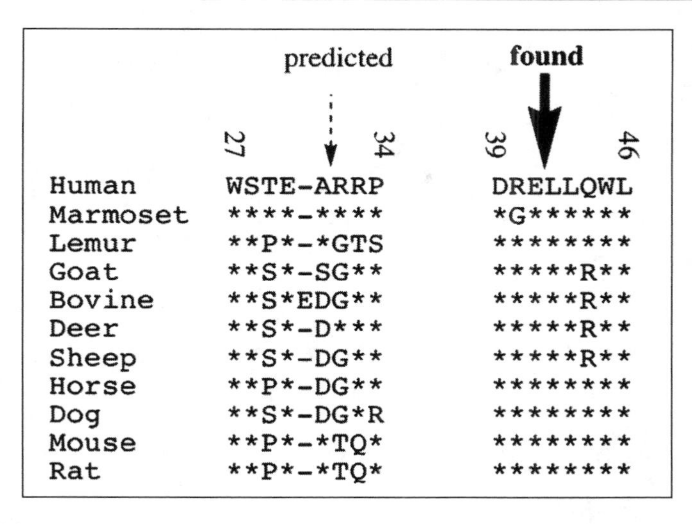

Figure 5. Pro-RLF conversion site at the B-chain connecting peptide junction: The bovine RLF protein sequence showed conversion occurs at glutamic acid in the conserved sequence DRELL extending the length of the mature B chain to 41 residues (numbering is based on the hRLF sequence. Based on cDNA deduced structures conversion was predicted to cleave at the pair of basic amino acid residues in human RLF leaving Ala B31 as C-terminal amino acid of the B chain. Such a pair of basic amino acid residues followed by proline is only present in human and marmoset. Sequence comparison shows high diversity within this region. Dashes signify insertions, asterisks indicate sequence identity (equine RLF partial sequence, accession AB033169).

of 0.6 ng/ml at day 39 (Fig. 4). The RLF receptor on the gubernaculum is present throughout development (Fig. 4).[31]

Since antiserum raised against human RLF does not cross-react with endogenous rat RLF, pregnant rats were injected (day-2, i.p.) with human RLF, sacrificed 30 min or 60 minutes after injection and the human RLF levels determined in the serum of the mother as well as the pups. It was noted that after 30 minutes the level of RLF was high in all rats tested. Rats that were sacrificed after 60 minutes however showed lower RLF-levels in the mother and female pups whereas RLF concentrations in male pups remained higher. From this experiment one can conclude that RLF passes the placental barrier and likely reaches the target tissues in the male pups and that high levels can only be attained by continuous administration of RLF. These results, clear as they were, do nothing to alleviate the anxiety created by the fact that exogenous RLF will not rescue the knock out phenotype. Did we not use the proper form of RLF?

A major disadvantage of cDNA-deduced structures is the uncertainty about the length of the individual peptide chains. Since the junction between the C terminus of the B-chain to the connecting peptide was not constant and in most species lacked the characteristic pair of basic amino acids (Fig. 5) one could not be sure about the conversion site. Isolation of RLF from bovine testis indeed showed a longer than usual B-chain.[33] The main component of the isolated material comprised 40 residues and a minor one of 45 residues (~20%). The extra five amino acids at the N terminus of the B chain suggested alternate processing by the signal peptidase. The C-terminal ends are identical so that the chain ends in glutamic acid followed by two leucines in the P1' and P2' position. This ELL-sequence is seen in all native RLFs (Fig. 5).[2,32,34-42] A comparison of testicular bovine RLF and synthetic bovine RLF, bearing a 31 residue B chain as deduced from the cDNA, showed that the longer, native structure had a reduced affinity to the mouse uterine RLF receptor.

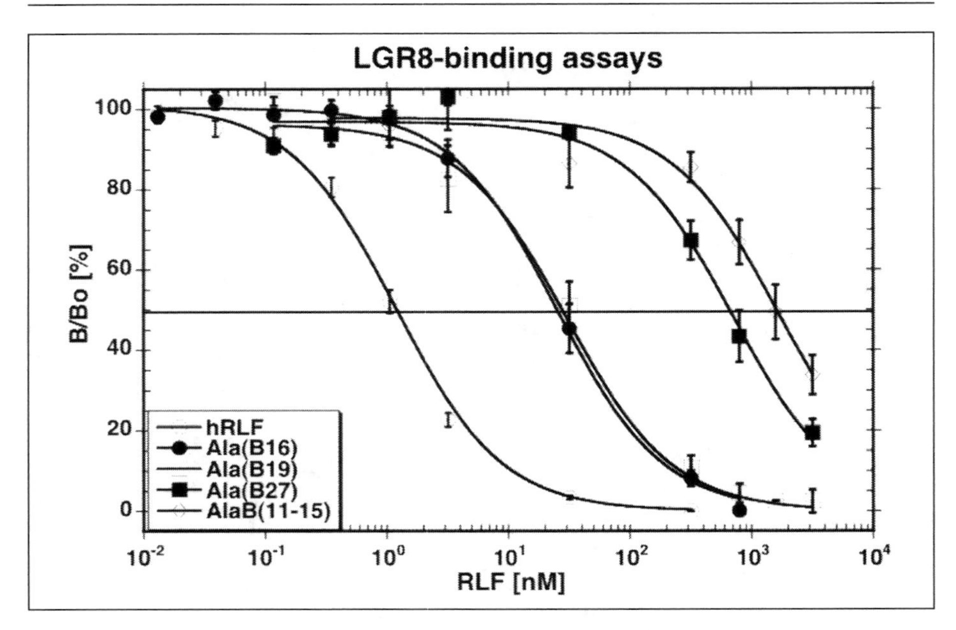

Figure 6. LGR8 binding human RLF and human RLF derivatives modified in the binding site.

This observation suggests that the C-terminal extension hampers receptor contact.[33] Of mouse RLF, however, neither the long- nor the short form was able to rescue the cryptorchid (-/-) mice.

Structural requirements for RLF-binding to LGR8 were explored by alanine scanning. The results showed clearly that the A-chain and the N-terminal ends of both chains did not contribute measurably to the binding energy. From earlier work it is known that replacement of tryptophan B27 by alanine reduces the affinity of RLF for the receptor by a factor of 200.[43] All residues beyond tryptophan could be deleted without loss of affinity if the carboxyl function is amidated.[44] Within the B chain two other residues appeared to be important for binding, namely arginine (B16) and valine (B19) (Fig. 6).[45,46] Although these three residues are present in human relaxin there are distinct differences between the hormones. Arginine in position B16 is located in the same relative position as Arg B17 in relaxin and valine in position B19 is equivalent to the relaxin receptor-binding residue Ile B20. Exchange of Arg B16 for Ala diminished binding only about 25-fold and the Val B19 Ala replacement caused an 18-fold loss of affinity. Even Gly A14 exchange for Ile, which nearly eliminated relaxin-binding to its receptor, reduced RLF-binding to LGR8 only to 53%.[45] It is noteworthy that the RLF B-chain alone binds LGR8 albeit marginally.[47]

Nothing discovered so far could explain the picomolar-binding constant associated with the interaction of RLF and LGR8. Each replacement in the B-chain gave a lukewarm response quite in contrast to the 1000-fold attenuation during the relaxin binding studies (Figs. 1,6). Comparing the receptor-binding sites of RLF and relaxin suggest that tryptophan B27 in RLF might be important for its cross-reactivity to LGR8. Porcine and human relaxins, for example, cross-react with LGR8 [8,48] while rat relaxin, which has glycine in place of tryptophan in the C-terminal region of the B-chain, does not.[48] On the other hand arginine (B13) in human relaxin is an LGR7 binding site while the equivalent position in RLF histidine (B12) can be replaced readily by alanine without reducing the binding affinity.[45,46] Interestingly, multiple replacements were not additive, for instance, individual replacement of residues B11 to B15 (GHHFV) in the mid-section of the B-chain made only small contributions to the RLF-binding, but when the whole block was exchanged for alanine (B11-15 Ala_5) the binding energy dropped about 1000-fold over wild-type which is 110 times the estimated additive effect of the exchange of the 5 residues individually.[45]

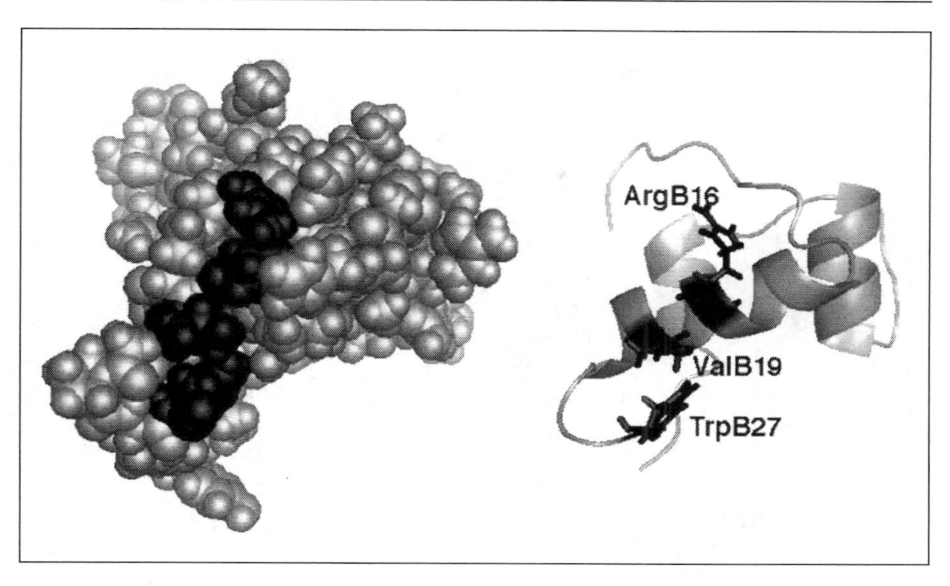

Figure 7. The 3 dimensional NMR structure of RLF displaying the residues Arg B1, Val B19 and Trp B27. For display the C-terminal flexible end B(28-31) was removed.

A similar study in which multiple surface residues of the B chain helix were replaced by alanine showed the same trend. When His B12 and Arg B16 were replaced by alanine the binding affinity was reduced by a factor of 33 compared to RLF, exceeding the additive effect by a factor 3[46] which is interesting but not significant when the overall affinity constant for RLF and LGR8 is in the nanomolar range.

The remarkable effect of the B11-15 Ala_5 raised questions concerning structure perturbations. Replacement of phenylalanine (B14) by alanine and to a lesser extend the replacement of glycine B11 by valine caused a change of the CD spectrum which was in part recovered by penta-alanine replacement of the B11-15 segment. This indicated that slight modification of the secondary structure is less important than the accumulative features supported by the side chains of histidines (B12, B13) and valine B15. The CD spectrum of the Ala (B12-15) showed only minor effects so that this region, pending evidence to the contrary, must be considered the major binding domain.[45]

Many of these findings were reinforced by an NMR structure of RLF published by Rosengren et al.[46] The indole ring of Trp B27 is proximate to the methyl groups of Leu B18 and Val B19. The chain beyond the C terminus of tryptophan is flexible and can easily move out of the way to free the indole group for receptor contact (Fig. 7). As suggested earlier the unique and constant GGP sequence (B23-25) is necessary to allow for the turn of the backbone. In fact, substitution of D-proline in position B25 limits the movement of B27Trp and decreases receptor binding.[43]

It has been suspected for some time that RLF is not as densely packed and well structured as other members of this protein family.[33,49] The slower tumbling of human RLF noted during NMR studies[46] clearly suggest a larger size in solution than that projected by the molecular mass. The probability that molecular motion could be a criterion in RLF receptor-binding led to the design of experiments that should enable one to trap the active conformation by immobilizing intra-molecular motion. A crosslink between the two C termini could be established by replacing arginine B26 with lysine and linking the side chain amino group through the Cterminus of the A chain using different length spacers (Fig. 8A). The study showed that binding activity was a function of the length of the spacer. These experiments were done with a B-chain ending in Trp

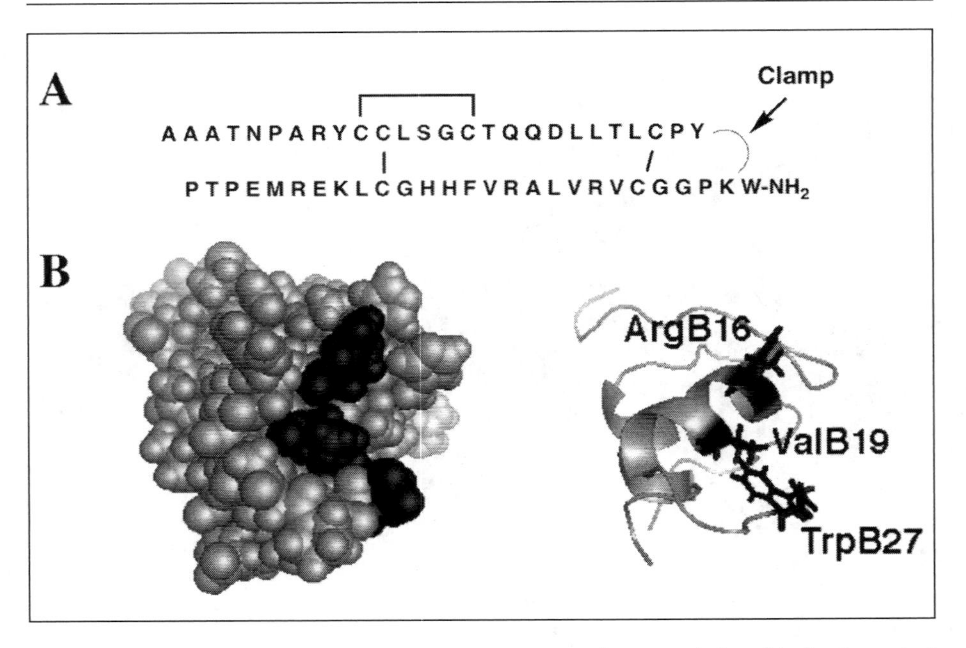

Figure 8. A) The structure of RLF derivatives bearing an intrachain crosslink within the C-terminal region of the B chain. The highlighted "clamp" was varied in length. Only the glycine containing crosslink led to full binding and activation of LGR8. B) Preliminary 3D structure of human RLF bearing a glycine spacer, Arg B16, Val B19 and Trp B27.

B27-amide.[44] Preliminary results of the NMR data reveal that the contact between the indole ring of Trp B27 and the mid-region of the B chain persists in this cross-linked derivative (Fig. 8B).

How large must RLF be to bind its receptor and initiate a signal? While the C-terminal truncation beyond a certain point interfered with LGR8 binding removing amino acids from the N terminus did not. In fact, the N-terminal region of the RLF A chain, which is devoid of any binding function, proved to be the seat of the signal initiation structure. Removal of the N-terminal end of the A chain up to the first cystine cross-link produced a tightly binding ligand that would not produce the signal to activate the cAMP cascade. Signal transmission is diminished when 7 residues are deleted from the N terminus and virtually disappears when eight or nine residues were removed from the N terminus of the A-chain. In subsequent experiments desA(1-8)RLF was used as RLFi (RLF-inhibitor), a specific potent competitive inhibitor of RLF action. The point of 50% inhibition of RLF by RLFi lies exactly at the 1:1 molar ratio of hormone and inhibitor when both concentrations are 2 ng/ml (Fig. 9).[50]

Searching for the signal initiation site RLF derivatives were synthesized with Arg A8 or Tyr A9 replaced by alanine. Both derivatives produced full cAMP activity meaning that the active region may be further down the chain or largely independent of the amino acid side-chain.

Unlike relaxin RLF has been clearly associated with a vital process such as the functional positioning of male gonads. The specific site of action is not known. The gubernaculum has garnered much attention as the dynamic tissue in this process, but again experiments with radioactive nucleosides and gubernaculum cells in culture gave only marginal results.[31] There is little doubt that RLF is indeed required for the process but the physiology of the process requires more attention before RLF can take its place in the clinic.

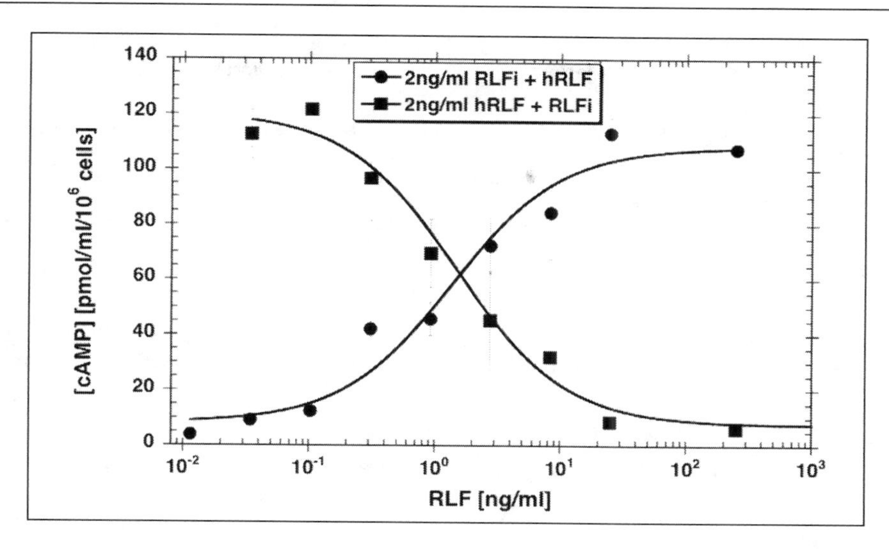

Figure 9. Inhibition assay of human RLF shortened at the N terminus of the A chain by 8 residues. Either 2 ng/ml of RLFi was added together with variable amounts of human RLF or 2 ng/ml human RLF was added together with variable amounts of RLFi. Each assay was performed in duplicate and two assays were pooled. The mean (±SEM) is presented in the graph.

References

1. Hisaw FL. Experimental relaxation of the pubic ligament of the guinea pig. Proc Soc Exp Biol Med 1926; 23:661-663.
2. Adham IM, Burkhardt E, Benahmed M et al. Cloning of a cDNA for a novel insulin-like peptide of the testicular Leydig cells. J Biol Chem 1993; 268:26668-26672.
3. Zimmermann S, Steding G, Emmen JM et al. Targeted disruption of the Insl3 gene causes bilateral cryptorchidism. Mol Endocrinol 1999; 13:681-691.
4. Nef S, Parada LF. Cryptorchidism in mice mutant for Insl.3 Nat Genet 1999; 22:295-299.
5. Overbeek PA, Gorlov IP, Sutherland RW et al. A transgenic insertion causing cryptorchidism in mice. Genesis 2001; 30:26-35.
6. Büllesbach EE, Schwabe C. A novel Leydig cell cDNA-derived protein is a relaxin-like factor (RLF). J Biol Chem 1995; 270:16011-16015.
7. Büllesbach EE, Schwabe C. Specific, high-affinity relaxin-like factor receptors. J Biol Chem 1999; 274:22354-22358.
8. Hsu SY, Nakabayashi K, Nishi S et al. Activation of orphan receptors by the hormone relaxin. Science 2002; 295:671-674.
9. Sherwood OD, O'Byrne EM. Purification and characterization of porcine relaxin. Arch Biochem Biophys 1974; 160:185-196.
10. Büllesbach EE, Schwabe C. Naturally occurring porcine relaxins and large-scale preparation of the B29 hormone. Biochemistry 1985; 24:7717-7722.
11. Canova-Davis E, Kessler TJ, Lee PJ et al. Use of recombinant DNA derived human relaxin to probe the structure of the native protein. Biochem 1991; 30:6006-6013.
12. Büllesbach EE, Schwabe C. Total synthesis of human relaxin and human relaxin derivatives by solid phase peptide synthesis and site-directed chain combination. J Biol Chem 1991; 266:10754-10761.
13. Schwabe C, Büllesbach E. Relaxin and the fine structure of proteins: R.G. Landers Co.; 1998.
14. Eigenbrot C, Randal M, Quan C et al. X-ray structure of human relaxin at 1.5 Å: Comparison to insulin and implications for receptor binding determinants. J Mol Biol 1991; 221:15-21.
15. Reinig JW, Daniel LN, Schwabe C et al. Isolation and characterization of relaxin from the sand tiger shark (odontaspis taurus). Endocrinology 1981; 109:537-543.
16. Büllesbach EE, Schwabe C. On the receptor binding site of relaxin. Int. J Peptide Protein Res 1988; 32:361-367.
17. Büllesbach EE, Schwabe C. The functional importance of the A chain loop in relaxin and insulin. J Biol Chem 1994; 269:13124-13128.

18. Büllesbach EE, Schwabe C. The relaxin receptor binding site geometry suggests a novel gripping mode of interaction. J Biol Chem 2000; 275:35276-35280.
19. Kobe B, Deisenhofer J. Crystal structure of porcine ribonuclease inhibitor, a protein with leucine-rich repeats. Nature 1993; 366:751-756.
20. Kobe B, Deisenhofer J. A structural basis of interactions between leucine-rich repeats and protein ligands. Nature 1995; 374:183-186.
21. Scott DJ, Layfield S, Yan Y et al. Characterization of novel splice variants of LGR7 and LGR8 reveals that receptor signaling is mediated by their unique LDLa modules. J Biol Chem 2006.
22. Büllesbach EE, Schwabe C. The trap-like relaxin-binding site of the leucine-rich G-protein-coupled receptor 7. J Biol Chem 2005; 280(14):14051-14056.
23. Brennand JE, Calder AA, Leitch CR et al. Recombinant human relaxin as a cervical ripening agent. Brit J Obst Gynaecol 1997; 104(7):775-780.
24. Dschietzig T, Stangl K. Relaxin: a pregnancy hormone as central player of body fluid and circular homeostasis. Cell Mol Life Sci 2003; 60:688-700.
25. Samuel CS, Unemori EN, Mookerjee I et al. Relaxin modulates cardiac fibroblast proliferation, differentiation and collagen production and reverses cardiac fibrosis in vivo. Endocrinology 2004; 145:4125-4133.
26. Seibold JR. Relaxin: lessons and limitations. Curr Rheumatol Rep 2002; 4:275-276.
27. Samuel CS, Zhao C, Bathgate RAD et al. Relaxin deficiency in mice is associated with an age-related progression of pulmonary fibrosis. FASEB J 2003; 17:121-.
28. Møller H, Prener H, Skakkebæk NE. Testicular cancer, cryptorchidism, inguinal hernia, testicular atrophy and genital malformations: Case control studies in Denmark. Cancer Causes Control 1996; 7:264-274.
29. John Radcliffe Hospital Cryptorchidism Study Group. Cryptorchidism: an apparent substantial increase since 1960. Br Med J 1986; 293:1401-1404.
30. Adham IM, Steding G, Thamm T et al. The overexpression of the INSL3 in female mice causes descent of the ovaries. Mol Endocrinology 2002; 16:244-252.
31. Boockfor FR, Fullbright G, Büllesbach EE et al. Relaxin-like factor (RLF) serum concentrations and gubernaculum RLF receptor-display in relation to pre- and neonatal development of rats. Reproduction 2001; 122:899-906.
32. Spiess AN, Balvert M, Tena-Sempere M et al. Structure and expression of the rat relaxin-like factor (RLF) gene. Mol Reprod Devel 1999; 54:319-325.
33. Büllesbach EE, Schwabe C. The primary structure and the disulfide links of the bovine relaxin-like factor (RLF). Biochemistry 2002; 41:274-281.
34. Burkhardt E, Adham IM, Hobohm U et al. A human cDNA coding for the Leydig insulin-like peptide (Ley I-L). Hum Genet 1994; 94:91-94.
35. Bathgate R, Balvers M, Hunt N et al. Relaxin-like factor gene is highly expressed in the bovine ovary of the cycle and pregnancy—Sequence and messenger ribonucleic acid analysis. Biol Reprod 1996; 55(6):1452-1457.
36. Pusch W, Balvers M, Ivell R. Molecular cloning and expression of the relaxin-like factor from the mouse testis. Endocrinology 1996; 137(7):3009-3013.
37. Roche PJ, Butkus A, Wintour EM et al. Structure and expression of Leydig insulin-like peptide mRNA in the sheep. Mol Cell Endocrinology 1996; 121(2):171-177.
38. Koskimies P, Spiess AN, Lahti P et al. The mouse relaxin-like factor gene and its promoter are located within the 3'region of the JAC3 genomic sequence. FEBS Lett 1997; 419:186-190.
39. Zimmermann S, Schottler P, Engel W et al. Mouse Leydig insulin-like (Ley I-L) gene—Structure and expression during testis and ovary development. Mol Reprod Develop 1997; 47(1):30-38.
40. Hombach-Klonisch S, Tetens F, Kauffold J et al. Molecular cloning and localization of caprine relaxin-like factor (RLF) mRNA within the goat testis. Mol Reprod Develop 1999; 53:135-141.
41. Hombach-Klonisch S, Kauffold J, Rautenberg T et al. Relaxin-like factor (RLF) mRNA expression in the fallow deer. Mol Cell Endocrinology 2000; 159:147-158.
42. Truong A, Bogatcheva NV, Schelling C et al. Isolation and expression analysis of the canine insulin-like factor 3 gene. Biol Reprod 2003; 69:1658-1664.
43. Büllesbach EE, Schwabe C. Tryptophan B27 in the relaxin-like factor (RLF) is crucial for RLF receptor binding. Biochemistry 1999; 38:3073-3078.
44. Büllesbach EE, Schwabe C. Synthetic cross-links arrest the C-terminal region of the relaxin-like factor (RLF) in an active conformation. Biochemistry 2004; 43:8021-8028.
45. Büllesbach EE, Schwabe C. The mode of interaction of the relaxin-like factor (RLF) with the leucine-rich repeat G protein-activated receptor 8. J Biol Chem 2006; 26136-26143.
46. Rosengren KJ, Zhang S, Lin F et al. Solution structure and characterization of the receptor binding surface of insulin-like peptide 3. J Biol Chem 2006; 28287-28295.

47. Del Borgo MP, Hughes RA, Bathgate RA et al. Analogs of insulin-like peptide 3 (INSL3) B-chain are LGR8 antagonists in vitro and in vivo. J Biol Chem 2006; 281(19):13068-13074.
48. Halls ML, Bond CP, Sudo S et al. Multiple binding sites revealed by interaction of relaxin family peptides with native and chimeric relaxin family peptide receptors 1 and 2 (LGR7 and LGR8). J Pharmacol Exp Ther 2005; 313(2):677-687.
49. Dawson NF, Tan YY, Macris M et al. Solid phase synthesis of ovine Leydig cell insulin-like peptide—a putative ovine relaxin? J Peptide Res 1999; 53:542-547.
50. Büllesbach EE, Schwabe C. LGR8 signal activation by the relaxin-like factor. J Biol Chem 2005; 280(15):14586-14590.

Chapter 3

Diverse Signalling Mechanisms Used by Relaxin in Natural Cells and Tissues:
The Evolution of a "Neohormone"

Richard Ivell,* Kee Heng and Ravinder Anand-Ivell

Abstract

The small peptide hormone relaxin is a member of a rapidly evolving family of hormones and growth factors, whose mode of action appears to be particularly adapted to purely mammalian physiology. It is representative of a new category of hormones, referred to as neohormones, which appear to have evolved specifically to accommodate the needs of viviparity, lactation and wound repair. The mechanism of receptor signalling has also evolved in this family, with older members using receptor tyrosine kinases and new members such as relaxin adopting 7-transmembrane G-protein coupled receptors. Although relaxin primarily generates cAMP as second messenger, studies of relaxin signalling show that this does not conform to a classic G-protein dependent activation of adenylate cyclase: it requires additional cytoplasmic components, it can involve further coupling to PI3-kinase and PKCζ and it is absolutely dependent on a tyrosine kinase activity linked closely to the relaxin receptor. Relaxin may also independently activate glucocorticoid receptors. This diversity of signalling leads to a broad range of possible downstream transcriptional effects. Finally, in tissues where relaxin is known to be effective, there is often also local relaxin induction, amplifying the effects of the endocrine hormone.

Introduction

The peptide hormone relaxin has evolved relatively recently with the advent of mammals, probably to address new functions essential to the success of this new order of vertebrates. These new functions include several strictly associated with viviparity, such as implantation, preparation of the birth canal and accommodation of increased blood/fluid volume during pregnancy. In addition, relaxin appears to be linked to other nonreproductive functions, which have in common their increased importance in postreproductive life, such as the antagonism of fibrosis, wound-healing and neoangiogenesis and vasodilation in the context of infarct situations. A selective advantage for postreproductive survival is only really seen amongst mammals, where older family members participate in child-rearing and family support. Together, this concept has allowed relaxin to be registered as the first bona fide member of what we have recently called the "neohormones",[1] a group of new molecules, which have recently evolved to address specifically mammalian physiology.

Relaxin, together with its male counterpart, insulin-like peptide 3 (INSL3), which evolved to facilitate the migration of the testis into the scrotum, both appear to have been derived from a common ancestor evident as the extant and highly conserved neuropeptide, relaxin-3.[2] Together

*Corresponding Author: Richard Ivell—Research Centre for Reproductive Health and School of Molecular and Biomedical Science, University of Adelaide, Adelaide, SA 5005, Australia. Email: richard.ivell@adelaide.edu.au

Relaxin and Related Peptides, edited by Alexander I. Agoulnik. ©2007 Landes Bioscience and Springer Science+Business Media.

with insulin, IGF1, IGF2, INSL4, INSL5 and INSL6, these small (ca. 6000 Dalton), structurally related peptides share in common a tripartite structure, with A-, B- and C(connecting)-domains, linked by two inter-domain and one intra-domain cystine bridges. A recent detailed bioinformatic study has shown how the relaxin-like branch of this family has arisen exclusively within the deuterostome lineage by a series of gene duplications.[3]

What is particularly interesting about the evolution of this family is that the extant members appear to have quite discrete physiological roles, with only very limited overlap in function. Furthermore, for some of the most recent members, such as INSL4 and INSL6, cumulative evidence to data suggests that they might not yet possess cognate receptors of high affinity. This relatively rapid acquisition of new functions and the evolution of relaxin as a neohormone appear in large part to be due to the family members having repeatedly co-opted novel proteins as receptors. Whilst the most ancient members (insulin, IGF1 and IGF2) all appear to employ single transmembrane domain tyrosine kinase receptors, which can form homo- and heterodimers amongst themselves and the orphan receptor IRR, it is noteworthy that one member of this group of peptides (IGF2) has already acquired an additional receptor in the form of the mannose-6-phosphate receptor. The oldest member of the relaxin-like subgroup, relaxin-3, in contrast, makes use of a 7-transmembrane domain G-protein-coupled receptor, GPCR135 (now RXFP3). The most closely related peptide to relaxin-3, INSL5, makes use of a related receptor, GPCR142 (now RXFP4). This is logical, since it appears that both these ligands and their receptors were subject to a genome-wide duplication event early in vertebrate evolution. Whilst relaxin-3 in mammals is largely a brain peptide, relaxin-3-like molecules from sub-mammalian vertebrates have also been found expressed in both testes and ovaries, just as has INSL5 in mammals. This is important, since it suggests that already early in vertebrate evolution there is an association between members of the relaxin-like family of peptide hormones and reproductive physiology. It can be presumed that this functionality would involve GPCR135-like receptors and it is interesting to observe that modern relaxin is also able to interact with this receptor, though not with GPCR142. What seems to have happened subsequently in evolution is that concomitant with the emergence of mammals, the development of viviparous young and a scrotal testis, further gene duplications gave rise first to a common relaxin/INSL3 ancestor and then to the two hormones themselves. These events were accompanied by the promiscuous acquisition by the relaxin/INSL3 ancestor of a new kind of GPCR, belonging to subclass C of these receptors, to which the receptors for LH and FSH also belong. This subclass of GPCR comprises a group of LGR (leucine-rich repeat-containing G-protein-coupled receptor) proteins, some of which are still orphans with no known ligand and one of which, LGR4, has been shown by gene ablation to have a marked role in reproductive development.[4] An assumed gene duplication event led to the establishment of LGR7 (now RXFP1) as the specific receptor for relaxin and LGR8 (now RXFP2) as the specific receptor for INSL3.

In summary, the evolution of the relaxin-like branch of this peptide family is characterized by a relatively rapid evolution (by gene duplication) and by the promiscuous acquisition of new receptor molecules of diverse classes. And this evolution is still ongoing, as witnessed by the existence of orphan members of both the ligand and the various cognate receptor families.

Evolution of Signalling Systems

All of the major classes and subclasses of signalling molecules are already encoded with the genome of the nematode *Caenorhabditis elegans*. Indeed, representatives of all of the major gene families are present already in yeast. Signalling systems evolved very early in cells in order to allow optimal interpretation of and hence responsiveness towards environmental information. Whilst originally this meant information about the uncontrolled extrinsic environment, the evolution of multicellularity implied the co-evolution of a controlled environment within the organism and hence of paracrine, juxtacrine and endocrine systems and later nervous system. It is clear therefore that when neohormone systems like that of relaxin and its receptor evolved, they did so because of strong selective pressure for the regulation of novel physiology and most importantly they evolved in an incremental fashion by modulating the existing information transduction systems. This latter concept means that for hormone systems, which have evolved recently, we should not be surprised to find a degree of redundancy in the signalling pathways involved. Nor need redundancy imply that the hormone system concerned is not essential. One of the key things that modern studies of postgenomic systems biology are teaching us is that we need to move away from regarding signal transduction in terms of monistic and qualitative single pathways, towards the holistic appraisal of a summation of multiple and quantitative networks of informational interactions.

Relaxin as a Model Neohormone System

The product of the relaxin (H2-relaxin in humans, relaxin-1 in other mammals) gene, as opposed to the brain peptide, relaxin-3, is involved in multiple and diverse physiological functions. Studies with knockout mice, in which either the peptide gene or that for its receptor, LGR7, have been ablated, show that only the preparation of the birth canal and nipple development are nonredundant actions of the hormone. The many other functions of relaxin, whilst redundant in the sense that there is no corresponding phenotype in knockout mice, are nevertheless very important and can be demonstrated clearly by, for example, the application of pure exogenous peptide into a living mammal.

This review will focus primarily on the immediate (<1 hr) properties of relaxin to activate signal transduction pathways inside cells. The longer term (>1 hr) effects of relaxin to induce a wide range of physiological effects will only be reviewed in as much as these aspects indirectly highlight the pleiotropic mechanisms of relaxin action and the obvious diversity of signalling systems that need to be addressed in order to achieve these long term actions. The multiple effects of relaxin seen in a large number of in vivo experiments in rodents, primates, as well as in a range of other mammals, can be summarized as follows:

- Tissue remodelling induced by the specific modulation of secreted metalloproteinase and TIMP expression, usually at the level of gene transcription. It is this property, which is largely responsible for the anti-fibrotic action of relaxin, for its wound-healing properties and for the remodelling of the birth canal (cervix, pubic symphysis).
- Neoangiogenesis caused by an induction of new blood vessels, partly linked to the tissue remodelling above and partly to the local induction of VEGF.
- Vasodilation resulting from increased local NO production. Part of this may be caused by the induction of iNOS in affected cells. All three of these listed properties synergize in wound-healing and in the relaxin response to infarct.
- Mesenchymal differentiation occurs in several reproductive tissues, e.g., the endometrial stroma and is linked to a cessation of proliferation and morphological changes (reflecting altered gene expression) associated with a postmitotic phenotype.
- Neurotransmitter effects are seen in the brain when relaxin is injected directly into specific brain regions. Whether relaxin is mimicking the action of local relaxin-3, or whether these effects are a natural consequence of relaxin in the brain is not yet clear.

All details leading to these different phenotypic actions are largely still uncharted. Whilst some effects can be explained by cell-specific expression of the LGR7 receptors, many would presuppose a complex network of signalling pathways being activated by relaxin to different extents in different cell types. A recent article exploring how many different promoter-reporter constructs can be activated by relaxin in a cell-line over-expressing LGR7 is a good and simple illustration of this phenomenon.[5] In an alternative approach, we are using microarrays to examine the diversity of genes being influenced by relaxin in cells naturally expressing the LGR7 receptor (unpublished).

Stimulation of cAMP Accumulation

The best described cellular response to relaxin is the induced elevation in intracellular and extracellular cAMP concentration.[6,7] This occurs within less than 1 min of exposure of cells to relaxin and it is a key feature of this effect for cells, which naturally express the LGR7 receptor, that cAMP levels remain elevated for at least 1 hour.[7] Two cell systems have been most commonly examined, the THP1 human monocytic cell-line and primary cultures of human endometrial stromal cells from the cycle (ESC cells). Whereas the cloning and heterologous expression of the LGR7 receptor shows that this can function as a typical Gs-coupled receptor able to activate adenylate cyclase, there are a number of features of the cells which naturally express relaxin receptors that indicate that relaxin signalling does not adhere to the simple G-protein coupled model.

- It is a feature of agents which induce G-protein coupled activation of adenylate cyclase that the cAMP generated becomes part of an acute negative feedback loop to restrict further cAMP generation. This has been shown for other systems to involve the activation by cAMP of protein kinase A (PKA) and subsequent phosphorylation and hence activation of phosphodiesterase (PDE) enzymes in the cell. Alternatively, PKA has been shown to induce the expression of some PDE gene transcripts,[8] or by phosphorylation even directly inhibit adenylate cyclase. Yet in all cell systems naturally expressing relaxin receptors, levels of cAMP remain elevated for a substantial time.
- All of the components involved in the ligand-dependent activation of cAMP via a stimulatory G protein normally reside within the plasma membrane and hence are active in the plasma membrane-containing fractions of disrupted cells. For THP1 cells, although the control stimulation by VIP showed that these components were intact and functional in the membrane preparations of the cells, relaxin failed to elicit any cAMP response, but appeared to require an additional cytoplasmic component to achieve a specific elevation of cAMP.[7]

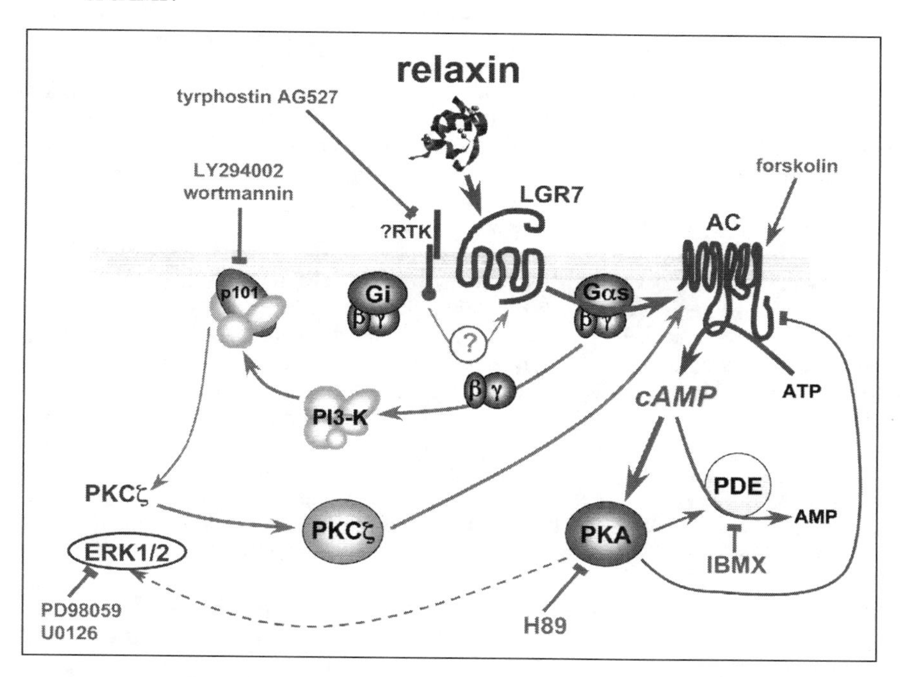

Figure 1. Cell signalling pathways shown to be activated by relaxin in cells which naturally express the relaxin receptor, LGR7. The role of a receptor tyrosine kinase (?RTK) is speculative and based on our studies using a variety of different tyrphostins. Our current investigations show that the phosphorylation of the MAP kinase ERK1/2 is a downstream consequence of PKA (protein kinase A) activation. (Other abbreviations: AC, adenylate cyclase; PKCς, protein kinase Cς; PI3-kinase, Phosphatidylinositol-3-kinase; PDE, phosphodiesterase; IBMX, the general PDE inhibitor, isobutyl-methyl-xanthanate). Inhibitory effects are indicated by transverse bars.

- The cell-specific increase in cAMP concentration caused by relaxin is inhibited completely by modest concentrations of a certain group of tyrphostins, but not by others.[7] Tyrphostins are highly specific inhibitors of tyrosine kinase activity and the subgroup of these inhibitors which are effective include ones (e.g., AG527, AG879) specific for the EGF and neurokinin membrane receptors. Other specific tyrphostins (e.g., AG1295) are without effect.[7] Although we have examined a large number of known tyrosine kinase proteins in an attempt to identify this important intermediate, as well as applying more generalized phosphoproteomic approaches, we have so far not been able to provide an unambiguous identification of this tyrphostin-sensitive component.

Our current thinking on the way that relaxin can activate different cell signalling pathways is summarized in Figure 1.

Relaxin Stimulation of PI3-Kinase and Protein Kinase Cς (PKCς)

It has been suggested in some studies using THP1 cells, or LGR7 over-expressing HEK293T cells, that activation of LGR7 by relaxin can also make use of Gi proteins.[9] In particular, it has been shown for these cells that some but not all (maximally 50%) of the cAMP generated by relaxin can be inhibited by application of the PI3-kinase inhibitor LY294002.[9,10] It is supposed that the βγ subunits released from Gi are able to activate PI3-kinase and that this in turn activates the atypical PKCς, which itself can stimulate adenylate cyclase allosterically by phosphorylation.[11] In an exhaustive set of studies, we have been able to show that this is not an essential part of the normal relaxin-stimulated cAMP generating machinery, but is a facultative mechanism which only comes into play once endogenous cAMP production exceeds about 3-fold over baseline level.[20] This latter situation rarely occurs in cells naturally expressing the relaxin receptor, unlike in LGR7 over-expressing cells[9,20] and usually requires the intervention of cAMP enhancing reagents such as forskolin or PDE inhibitors (unpublished). Nevertheless, an older study does suggest that one of the consequences of relaxin action in ESC cells may indeed be the translocation of PKC to the plasma membrane,[12] though studies using the inhibitor wortmannin suggest that this might not necessarily be linked to the activation of adenylate cyclase.[13]

Is Relaxin Action Connected to Ca^{++} Flux in Stimulated Cells?

Given that relaxin is able to activate PKC in some circumstances and that it might be able to activate PI3-kinase, it seems plausible that relaxin could make use of intracellular Ca^{++} as an additional second messenger. Indeed, in an older article[14] evidence is presented that relaxin is able to induce Ca^{++} transients in primary cultures of human granulosa-lutein cells. We have carried out extensive studies to explore a possible role for intracellular Ca^{++} in relaxin signalling in THP1 monocytes (Fig. 2). In spite of the correct functioning of all positive and negative controls, we were unable to detect any influence of relaxin on intracellular Ca^{++} in THP1 cells as monitored by changes in Fluo-3AM fluorescence (Fig. 2). Furthermore, we have been unable to detect any effect on relaxin-dependent cAMP generation of the intracellular calcium chelator BAPTA/AM or the PKC inhibitor bisindoylmaleimide, as well as of the external application of the Ca^{++} chelating agent EGTA (unpublished).

Relaxin Can Act Through the Glucocorticoid Receptor (GR)

One of the most exciting recent discoveries in the area of relaxin signalling was the discovery that relaxin in some as yet unknown way is able to directly activate the glucocorticoid receptor (GR) inside the cell (THP1 cells), so that it is then able specifically to activate or inhibit known glucocorticoid-dependent genes via a classic GR-responsive element.[15] This effect of relaxin on a GR-dependent gene promoter was recently confirmed for heterologously transfected LGR7 over-expressing HEK293T cells,[5] where this effect was shown to be partly dependent on the presence of the LGR7 receptor. We have recently shown that this effect is unlikely to be linked to the LGR7-dependent activation of adenylate cyclase, since neither the specific inhibitor of GR, RU486, nor the GR agonist dexamethasone, have any effect on relaxin-dependent cAMP accumulation in THP1 cells.[20]

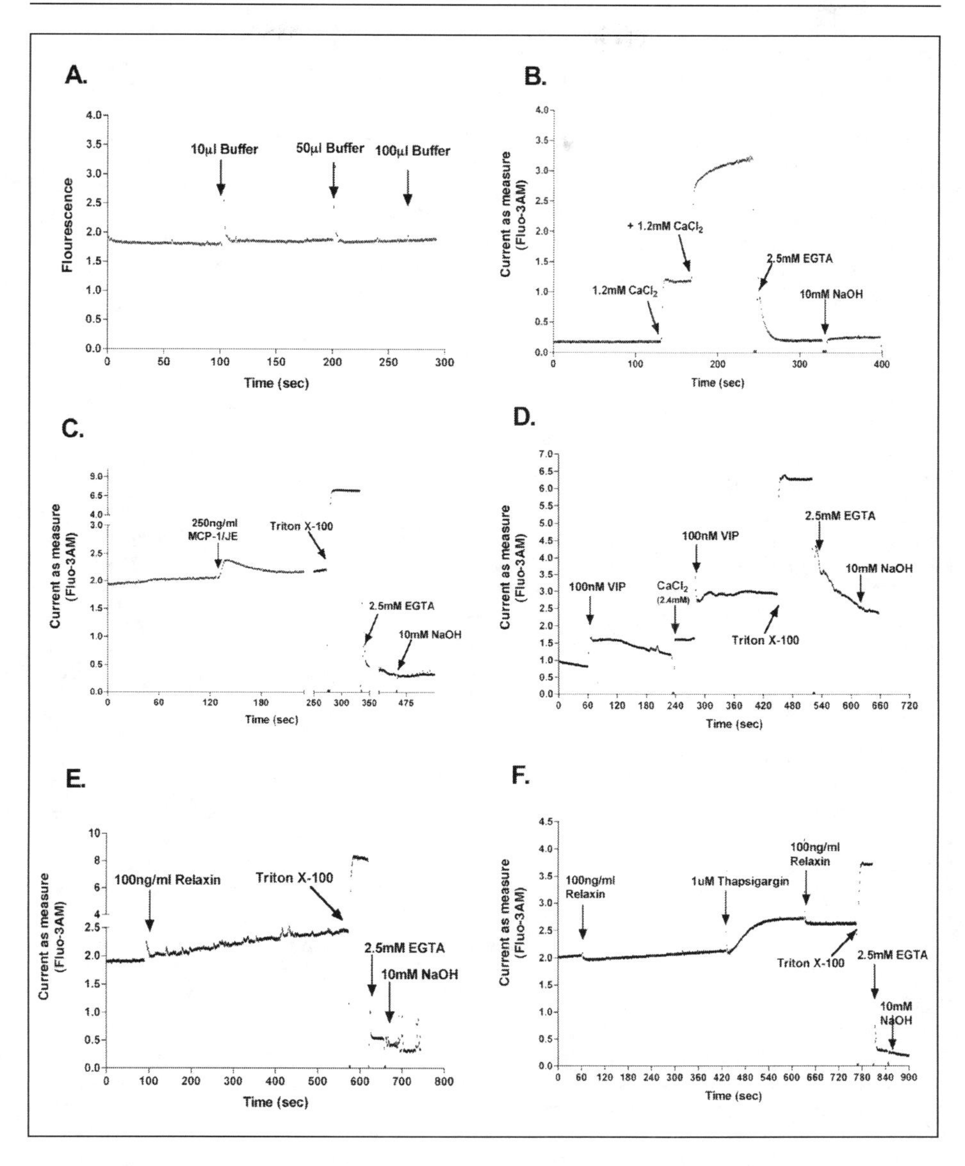

Figure 2. Lack of involvement of intracellular Ca^{++} flux in the relaxin signalling pathways operating in human THP1 monocytes. THP1 cells were loaded with the Ca^{++} chelating fluorophore Fluo-3AM and changes in fluorescence measured in a cuvette system upon the addition of various effectors. A) Control to check that additional volume per se was without effect. B) The efficacy of the fluorophore was checked by adding calcium as indicated. (C-F) At the indicated time points (arrows) various effectors were added to the cuvettes and the fluorescence emission at 526 nm measured (C, the chemokine analogue MCP1/JE; D, VIP; E, relaxin; F, relaxin with and without thapsigargin treatment). To terminate all reactions, Triton X-100 was added to liberate all of the calcium from the intracellular stores. Then, the reactions were quenched with EGTA and finally, by the addition of NaOH solution.

Downstream Effects of Relaxin-Dependent Cell Signalling

As mentioned above, many of the phenotypic consequences of relaxin action become evident after longer term stimulation by the hormone. These effects invariably involve genomic effects caused by the activation or suppression of specific genes. Whilst for some genes the mechanism of action is simple to reconstruct since the promoter regions may contain simple response elements known to be downstream of activated PKA (e.g., CREB responsive elements, as in the promoter of the VEGF gene), for other genes the promoters do not yet reveal such simple traits. In ESC cells, it is known that some of the genes induced by relaxin are those that can be also induced by cAMP in the general decidualization phenotype, e.g., prolactin or IGF-BP1. It is known that the gene for the NO generating enzyme iNOS can also be induced by cAMP. However, other genes may be specifically induced by relaxin, which are not normally associated with elevated cAMP. Further studies are needed to clarify the extent to which other pathways involved in gene expression are activated by relaxin, or which require relaxin stimulation as part of a more complex synergistic mechanism.

Other downstream mechanisms, which do not appear to involve new gene expression, mostly appear to involve PKA. So for example, we have recently shown that the activation of the MAP kinase pathway and hence its downstream growth-promoting properties, can be inhibited by treatment of cells with the specific PKA inhibitor H89 (unpublished). The extensive and elegant studies of Barbara Sanborn and colleagues have shown that the ability of relaxin to induce quiescence in oxytocin-induced or spontaneously contractile rat myometrial cells is largely due to a PKA-dependent recruitment of inhibitory components to an AKAP protein on the plasma membrane.[16] However, these studies were largely performed using rat cells where the quiescent effect of relaxin is very marked. Curiously, in the human myometrium, whilst there is good evidence for the expression of LGR7 receptors,[17] the ability of relaxin to suppress oxytocin-induced or spontaneous contractility cannot be demonstrated.[18] This would suggest that not all downstream mechanisms are equivalent in different cells and tissues.

Local Relaxin Systems

Cell signalling is usually looked at in the context of monolayer cells growing in culture media and it is assumed that this reflects a situation akin to that of a freely circulating hormone impacting on cells within a tissue. However, for many growth factors and hormones, this is rarely the case. Especially for relaxin, we know that many of the tissues, which are responsive to relaxin (i.e., express functional LGR7), also express relaxin (or some surrogate, such as immunological relaxin epitopes, mRNA-derived PCR products, etc.,). This is particularly relevant for the physiology, which is of interest to us: the role of relaxin to induce the differentiation of the endometrial stroma, a process known as decidualization. We have shown that relaxin, largely through the induction of cAMP in proliferating and undifferentiated human endometrial stromal cells, causes these to change morphology and gene expression pattern to those of the typical secretory decidual cells of the latter half of the human menstrual cycle.[8] In vivo this process is accompanied by an induction of secretory activity in the juxtaposed glandular epithelial cells, which includes an upregulation of relaxin expression within the epithelial cells, at least in humans and primates.[19] What this means is that once the endometrial differentiation process is initiated, by whatever hormonal agent (e.g., steroids, relaxin, prostaglandins), then local relaxin production would generate a feed-forward loop to maintain high local cAMP in the tissue and hence the advanced differentiation status characteristic of the second half of the menstrual cycle and early pregnancy. Similar situations are probably occurring also in other tissues, where both hormone and receptors are shown to be expressed, including in tumors.

Concluding Remarks

The peptide relaxin superficially appears to be just like many other GPCRs responding to ligand by activating adenylate cyclase in a conventional Gs-protein coupled fashion. However, where receptors are expressed naturally and at relatively low frequency, this production of cAMP appears to involve additional cytoplasmic components, of which one at least comprises a tyrosine kinase activity. This increased level of signalling complexity is reflected also in the pleiotropy of effects seen at the level of gene expression and downstream physiological responses, where it is clear that the pattern of genes, which respond to relaxin, reflects strongly the evolution of this neohormone. Thus, the unravelling of the signalling pathways used by relaxin in achieving its specific cell responses will provide important insights into both the mechanisms by which this neohormone has evolved, as well as the manner in which cells can use signalling complexity to encode higher order levels of response specificity. Relaxin, as the holotypic member of this new group of neohormones, offers an important model system with which to analyse the way that cells process incoming biological information.

Acknowledgements

The authors gratefully acknowledge financial support from the National Health and Medical Research Council of Australia, BioInnovation SA, Adelaide.

References

1. Ivell R, Bathgate RAD. Hypothesis: Neohormone systems as exciting targets for drug development. Trends Endocrinol Metab 2006; 17:123.
2. Bathgate RA, Hsueh AJ, Ivell R et al. International Union of Pharmacology. Recommendations for the nomenclature of receptors for relaxin family peptides. Pharmacol Rev 2006; 58:7-31.
3. Wilkinson TN, Speed TP, Tregear GW et al. Evolution of the relaxin-like peptide family. BMC Evol Biol 2005; 5:14.
4. Mazerbourg S, Bouley DM, Sudo S et al. Leucine-rich repeat-containing G protein-coupled receptor 4 null mice exhibit intrauterine growth retardation associated with embryonic and perinatal lethality. Mol Endocrinol 2004; 18:2241-2254.
5. Halls ML, Bathgate RA, Summers RJ. Comparison of signalling pathways activated by the relaxin family peptide receptors, RXFP1 and RXFP2, revealed using reporter genes. J Pharmacol Exp Ther 2006; (epub ahead of print).
6. Fei DT, Gross MC, Lofgren JL et al. 1990. Cyclic AMP response to recombinant human relaxin by cultured human endometrial cells—a specific and high throughput in vitro bioassay. Biochem Biophys Res Commun 1990; 170:214-222.
7. Bartsch O, Bartlick B, Ivell R. Relaxin signalling links tyrosine phosphorylation to phosphodiesterase and adenylyl cyclase activity. Mol Hum Reprod 2001; 7:799-809.
8. Bartsch O, Bartlick B, Ivell R. Phosphodiesterase 4 inhibition synergizes with relaxin signalling to promote decidualization of human endometrial stromal cells. J Clin Endocrinol Metab 2004; 89:324-334.
9. Halls ML, Bathgate RA, Summers RJ. Relaxin family peptide receptors RXFP1 and RXFP2 modulate cAMP signalling by distinct mechanisms. Mol Pharmacol 2006; 70:214-226.
10. Nguyen BT, Yang L, Sanborn BM et al. Phosphoinositide 3-kinase activity is required for biphasic stimulation of cyclic adenosine 3',5'-monophosphate by relaxin. Mol Endocrinol 2003; 17:1075-1084.
11. Nguyen BT, Dessauer CW. Relaxin stimulates protein kinase C zeta translocation: requirement for cyclic adenosine 3',5'-monophosphate production. Mol Endocrinol 2005; 19:1012-1023.
12. Kalbag SS, Roginsky MS, Jelveh Z et al. Phorbol ester, prolactin and relaxin cause translocation of protein kinase C from cytosol to membranes in human endometrial cells. Biochim Biophys Acta 1991; 1094:85-91.
13. Bartsch O, Bartlick B, Ivell R. Relaxin signal transduction couples tyrosine phosphorylation to cAMP upregulation. In: Tregear GW, Ivell R, Bathgate RAD, Wade JD, eds. Relaxin 2000: Proceedings of the 3rd International Conference on Relaxin and Related Peptides. Kluwer Academic Publishers; Dordrecht, Netherlands 2001; 309-316.
14. Mayerhofer A, Engling R, Stecher B et al. Relaxin triggers calcium transients in human granulosa-lutein cells. Eur J Endocrinol 1995; 132:507-513.
15. Dschietzig T, Bartsch C, Stangl V et al. Identification of the pregnancy hormone relaxin as glucocorticoid receptor agonist. FASEB J 2004; 18:1536-1538.
16. Dodge KL, Sanborn BM. Evidence for inhibition by protein kinase A of receptor/G alpha(q)/phospholipase C (PLC) coupling by a mechanism not involving PLCbeta2. Endocrinology 1998; 139:2265-2271.
17. Ivell R, Balvers M, Pohnke Y et al. Immunoexpression of the relaxin receptor LGR7 in breast and uterine tissues of humans and primates. Reprod Biol Endocrinol 2003; 1:114.
18. Downing SJ, Hollingsworth M. Action of relaxin on uterine contraction—a review. J Reprod Fertil 1993; 99:275-282.
19. Ivell R, Einspanier A. Relaxin peptides are new global players. Trends Endocrinol Metab 2002; 13:343-348.
20. Anand-Ivell R, Heng K, Ivell R. Relaxin signaling in THP-1 cells uses a novel phospholyrosine-dependent pathway. Mol Cell Endocrinol 2007; (in press).

Chapter 4

Relaxin Physiology in the Female Reproductive Tract during Pregnancy

Laura J. Parry* and Lenka A. Vodstrcil

Abstract

The characteristic functions of relaxin are associated with female reproductive tract physiology. These include the regulation of biochemical processes involved in remodeling the extracellular matrix of the cervix and vagina during pregnancy and rupture of the fetal membranes at term. Such modifications enable the young to move unimpeded through the birth canal and prevent dystocia. However, relaxin's physiological actions are not limited to late gestation. New functions for this peptide hormone in implantation and placentation are also emerging. Relaxin promotes uterine and placental growth and influences vascular development and proliferation in the endometrium. This chapter provides an overview of the current literature on relaxin physiology in the uterus, cervix and vagina of pregnant females and the impact on fetal health. It also outlines the potential mechanisms of relaxin action, particularly in the cervical extracellular matrix and uterine endometrium.

Introduction

A role for the peptide hormone relaxin in female reproductive tract physiology was first described by Hisaw in 1926. He injected serum from pregnant guinea pigs into virgin guinea pigs and observed a relaxation of the pubic symphysis.[1] This allows the two innominate bones to widen and facilitate passage of the fetus with its relatively large head through the pelvic girdle. These experiments generated the concept that relaxin was an important hormone for successful parturition and delivery of live young. It was not until the 1970s, when highly purified porcine and rat relaxin became available, that experiments could begin to investigate the physiological effects of relaxin treatment in the female reproductive tract (reviewed in ref. 2). Many studies used ovariectomized pigs and rats, with hormone replacement paradigms. In the 1980s, Sherwood and colleagues developed a monoclonal antibody to rat relaxin (MCA1) and used it to neutralize endogenous relaxin in pregnant rats. Both these approaches showed that in the absence of relaxin there were substantial delays in the onset of labor, prolonged duration of delivery and a greater incidence of neonate mortality at birth.[3-5] Relaxin gene knockout mice ($Rln1^{-/-}$) were developed by Zhao and colleagues in 1999,[6] and provided an equally valuable tool to further investigate the role of relaxin in reproductive tract physiology. The $Rln1^{-/-}$ females give birth to live young without any apparent sign of dystocia, despite having abnormal cervical and vaginal morphology and no elongation of the pubic symphysis.[6] This illustrates the complexity in the study of relaxin physiology. There is considerable variation in the sources and secretion of relaxin during pregnancy, as well as the localization of the receptors for relaxin. Many actions of relaxin are specific to certain species and to date, there are no obvious clinical conditions in pregnant women associated with

*Corresponding Author: Laura Parry—Department of Zoology, University of Melbourne, Parkville, Victoria, 3010, Australia. Email: ljparry@unimelb.edu.au

Relaxin and Related Peptides, edited by Alexander I. Agoulnik. ©2007 Landes Bioscience and Springer Science+Business Media.

relaxin deficiency.[7,8] However, relaxin treatment in pregnant animals clearly has several effects that could be perceived as beneficial for a successful pregnancy as well as facilitating the process of labor. This review outlines the current literature on relaxin physiology in the cervix, vagina and uterus of pregnant females, with an emphasis on the potential mechanisms of relaxin action in each tissue.

Relaxin Receptors

The receptor for relaxin is a leucine-rich repeat containing, heterotrimeric guanine nucleotide binding (G-protein)-coupled receptor (GPCR) known as LGR7,[9-11] recently assigned the nomenclature RXFP1.[12] Activation of LGR7 by its ligand stimulates a G_s-cAMP-protein kinase A-dependent signaling pathway.[13,14] Localization of relaxin receptors in the female reproductive tract initially relied on biotinylated porcine relaxin and radiolabeled ligand. Relaxin binding sites were found mainly in epithelial cells in the cervix and vagina of rats and pigs,[15,16] and the myometrium of the rat uterus.[17-19] Once the human, mouse and rat relaxin receptors (LGR7/Lgr7) were cloned,[9,11,20] researchers were able to analyze the expression and localization of Lgr7 more accurately. In the mouse, staining for Lgr7-specific β-galactosidase activity was identified underneath the basal layer of the vaginal epithelium and in the circular layer of the myometrium.[21,22] We have shown that Lgr7 mRNA is predominantly expressed in the myometrium compared with the endometrium or placenta of pregnant mice, with a down-regulation in myometrial Lgr7 at term.[23,24] The receptor is also highly expressed in the cervix of pregnant mice, but surprisingly, there is no surge in Lgr7 mRNA concentrations in the later stages of gestation to coincide with the dramatic changes in stromal extracellular matrix remodeling. Using in situ hybridization, we localized Lgr7 mRNA in the mouse cervix and vagina to the stromal tissue underlying the basal layer of epithelial cells (Fig. 1). Lgr7 was not highly expressed within the epithelium, which contradicts data from previous studies in the rat and pig.[15,16] However, the luminal epithelium is the predominant cell type in the human cervix and vagina (obtained from premenopausal hysterectomy patients) that expresses relaxin binding sites.[25] Relaxin also binds to the circular and longitudinal smooth muscle layers and vascular smooth muscle cells associated with blood vessels in the human cervix and vagina.[25]

The pattern of uterine LGR7 expression in primates and humans is very similar to the cervix and vagina but there is no consensus between studies on the predominant region expressing relaxin receptors. Immunoreactive LGR7 has been localized to the endometrial stromal and epithelial compartments in human and marmoset monkey uterus,[26,27] with more intense staining in the secretory phase of the cycle. Autoradiography studies contradict this finding and show ^{32}P-labeled human relaxin predominantly in the glandular and luminal epithelium (Fig. 2A in ref. 28). Furthermore, LGR7 mRNA expression is significantly higher in isolated human glandular epithelial and decidual cells compared with endometrial stromal cells.[29] Localization of biotinylated porcine relaxin in the uterus of hysterectomized women was similarly restricted to luminal and glandular epithelial cells,[25] although a few cells in the stromal extracellular matrix were also positive for relaxin binding sites. Both human relaxin binding sites and LGR7 mRNA in the endometrium are markedly up-regulated during the early secretory phase of the cycle (Fig. 2B in ref. 28). Only one study has identified LGR7 expression in human cultured myometrial cells,[26] although biotinylated porcine relaxin binding was shown in the myometrium of the marmoset monkey[30] and human.[25] Another proven target tissue for relaxin action in humans is the fetal membranes of late gestation. Initial studies with biotinylated porcine relaxin revealed prominent labeling in the amnion epithelial cells and placental villi projecting into the lacuna system, shown at high magnification to be in the syncytiotrophoblast cells.[25] The recent work of Lowndes et al[31] demonstrated specific LGR7 gene transcripts and immunoreactive LGR7 in the human decidua and chorionic cytotrophoblast, with very low expression of the receptor in the amniotic epithelium. The differences between studies may be explained by the variety of techniques used and issues associated with specificity of antibodies, biotinylated molecules and radiolabeled ligands, although all studies cited include appropriate negative controls. Another explanation is that these discrepancies may be due to the expression of various LGR7 splice variants which may or may not be functional.[32] Future LGR7 localization studies should include concurrent gene and protein analysis of the known full-length functional LGR7 variant to resolve the questions related to temporal and spatial expression of these receptors in the human uterus throughout the cycle and during pregnancy.

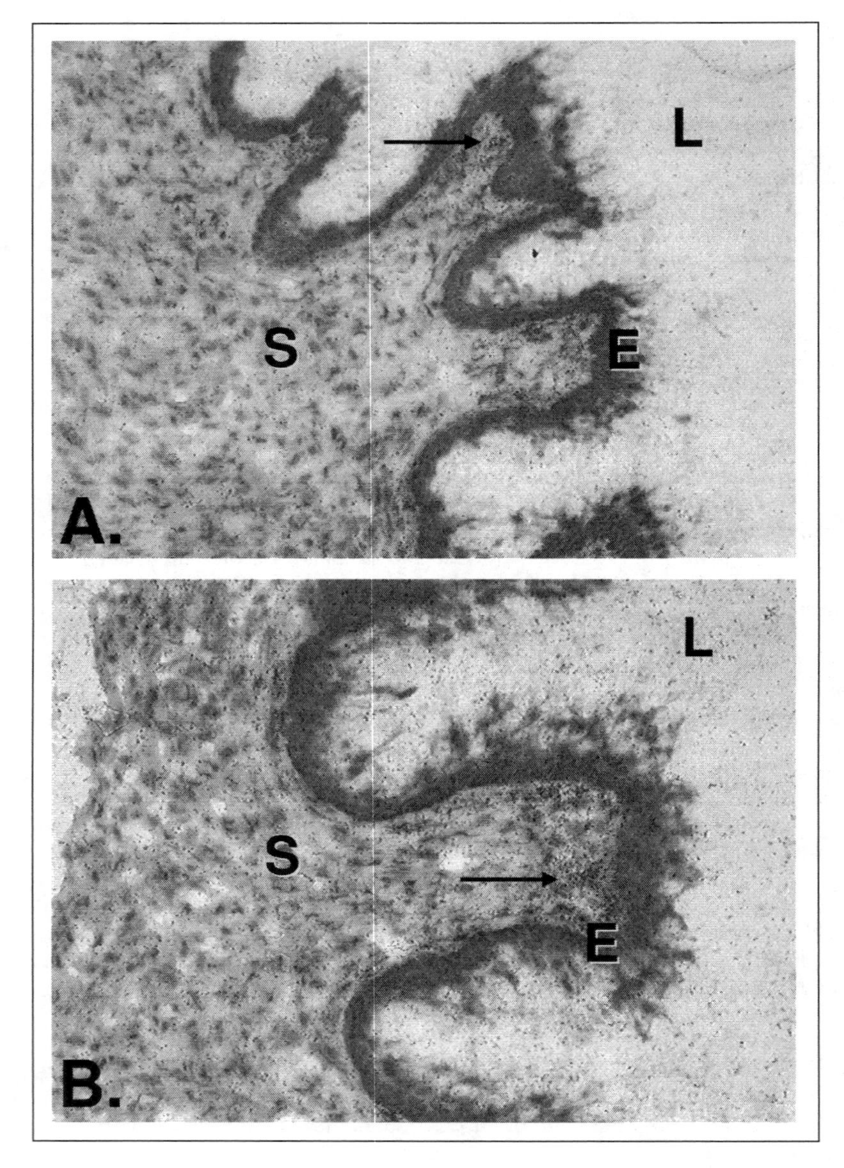

Figure 1. Localization of Lgr7 mRNA by in situ hybridization in the cervix (A) and vagina (B) of Rln1$^{+/+}$ mice on day 14.5 gestation. Positive hybridization signals (arrow) are present in the stromal tissue underlying the basal layer of epithelial cells. S, stroma; E, epithelium; L, lumen.

Effects of Relaxin on the Cervix

Several functional studies in rodents and pigs have demonstrated an important role for relaxin in the progressive softening of the cervix in the second half of gestation. There is a substantial reduction in cervical wet weight and extensibility in relaxin-deficient rats, with a higher incidence of neonate mortality.[3-5,33,34] Rln1$^{-/-}$ and Lgr7$^{-/-}$ mice also have abnormal cervical morphology but are able to give birth to live young.[6,21,22] The most obvious cervical phenotypes in pregnant Rln1$^{-/-}$ mice are the increased density of stromal extracellular matrix (particularly collagen) and a lack of epithelial proliferation (Fig. 3). These phenotypes are reversed by treatment with exogenous relaxin.[35] Morphological changes in the cervix induced by relaxin treatment are the increased area of luminal

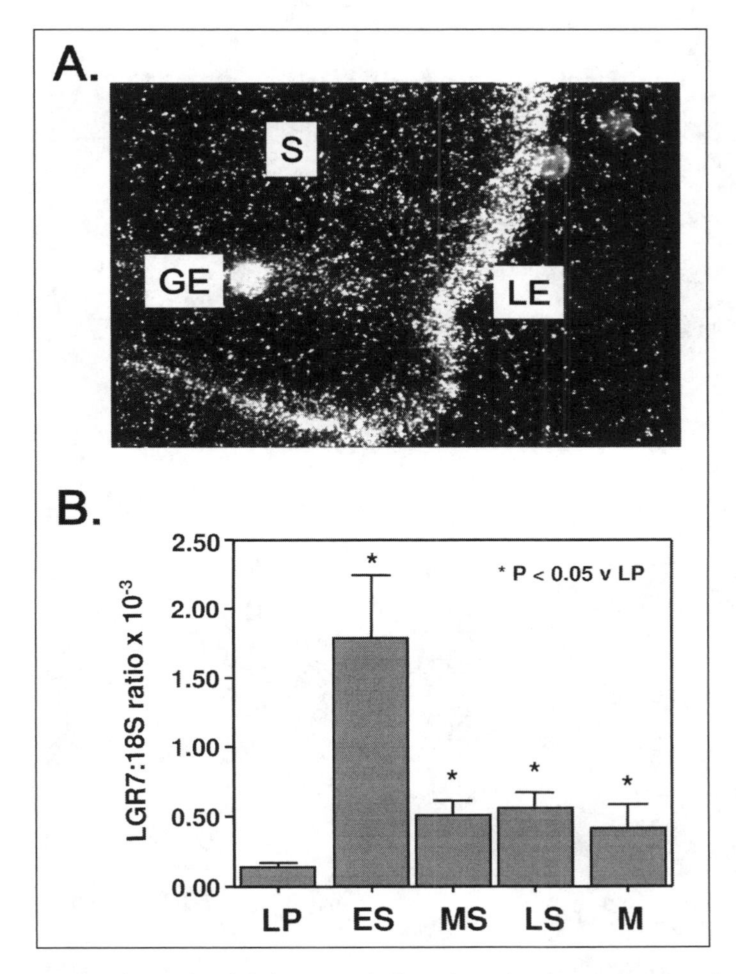

Figure 2. A) Autoradiographic localization of [^{33}P]-relaxin binding sites using photographic emulsion on a slide mounted section of human uterus obtained during the early to mid secretory phase. The dark field image shows silver grains in the glandular (GE) and luminal epithelial (LE) cells of the endometrium. B) LGR7 mRNA concentrations in the endometrium at five phases of the menstrual cycle. EP: early proliferative, ES: early secretory, MS: mid secretory, LS: late secretory, M: menstrual phase of the cycle. Adapted with permission from C.P. Bond et al. J Clin Endocr Metab; 89:3477-3485. ©2004 The Endocrine Society.

involutions and dispersal of collagen fibers, particularly in these involutions. There is also a marked proliferation of the epithelium, greater numbers of vacuolated epithelial cells and a large increase in the amount of a mucopolysaccharide lining the epithelium.[35] Similar findings were described in the ovariectomized rat and pig relaxin-replacement models, but in addition, relaxin increased the percentage hydration and glycosaminoglycan content in the cervix.[36,37]

Current theories on the mechanisms of relaxin action in the cervix focus predominantly on collagen dispersal and/or degradation. Relaxin treatment has little effect on cervical collagen content in pregnant ovariectomized rats or pigs.[36,38] Similarly, there is no difference in the percent-

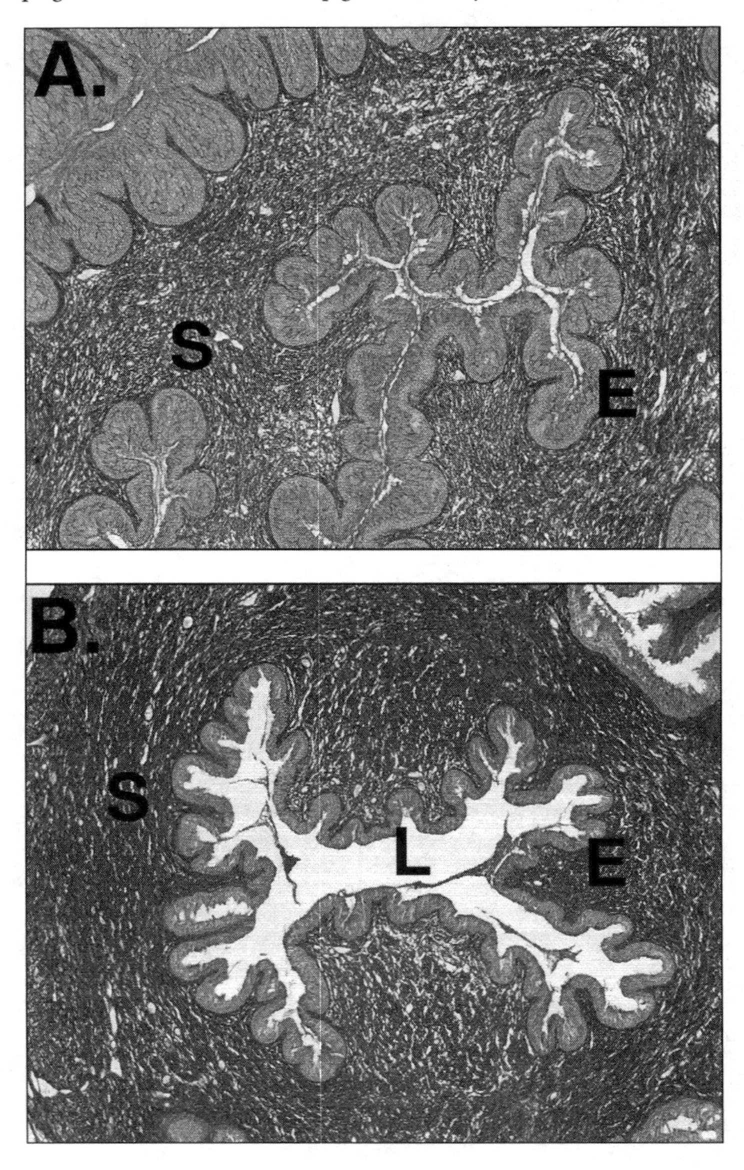

Figure 3. Collagen fibre density in the cervix of pregnant (A) $Rln1^{+/+}$ and (B) $Rln1^{-/-}$ mice on day 16.5 gestation. S, stroma; E, epithelium; L, lumen.

Table 1. The in vivo effects of relaxin deficiency and relaxin treatment on collagen and matrix metalloproteinase (MMP) gene expression in the cervix of late pregnant $Rln1^{-/-}$ mice

	$Rln1^{+/+}$		$Rln1^{-/-}$ v $Rln1^{+/+}$	$Rln1^{-/-}$ + Relaxin
	Gestation	**Term**		
Collagen I	↑	↓	↑ (day 18.5 only)	↓ on d18.5
Collagen III	↑	↓	↑ (day 18.5 only)	*Not measured*
TGFβ-1	↑	↓	↔	↔
MMP-13	↓	↓	↑ (highest at term)	↓ on d18.5
MMP-3	↑	↔	↑ (day 18.5 only)	*Not measured*
MMP-7	↓	↓	↑	*Not measured*
MMP-9	↓	↓	↑	*Not measured*
MMP-2	↑	↑	↔	↔

age hydroxyproline content in the cervix of late pregnant $Rln1^{+/+}$ and $Rln1^{-/-}$ mice.[39] These data are explained, in part, by the increased expression of α_1(I) and α_1(III) collagen in the cervix throughout gestation in $Rln1^{+/+}$ and $Rln1^{-/-}$ mice. The only difference between the genotypes was observed on day 18.5 gestation when α_1(I) collagen mRNA levels decrease significantly in $Rln1^{+/+}$ mice but remain high in $Rln1^{-/-}$ mice.[35] These data suggest that de novo collagen is synthesized in increasing quantities in the cervix throughout gestation and that a lack of relaxin does not result in abnormally high amounts of collagen in these tissues. However, administration of human relaxin to pregnant $Rln1^{-/-}$ mice caused a significant decrease in α_I(I) collagen gene expression in the cervix on day 18.5 gestation,[35] demonstrating the ability of exogenous relaxin to reduce collagen synthesis. Relaxin may, therefore, be capable of suppressing collagen synthesis by a direct action on cervical fibroblasts.

The other common hypothesis of relaxin action involves activation of collagen degrading enzymes in the extracellular matrix. Relaxin stimulates matrix metalloproteinase (MMP)-1 activity in cultured guinea pig cervical cells,[40] and MMP-1, gelatinases and stromelysin-1 in human lower uterine segment fibroblasts[41] and normal human cervical stromal cells.[42] These results using in vitro cell culture systems demonstrated that relaxin is a positive regulator of MMP expression. But in the cervix of wild type mice, there is no correlation between increased MMP expression and changes in tissue architecture. Only the gelatinase MMP-2 mRNA levels are greater at term compared with earlier stages of gestation. There are significant decreases in MMP-13 and MMP-7 expression and no change in MMP-9 and MMP-3. The situation in pregnant $Rln1^{-/-}$ mice is reversed. Expression of all MMPs examined, except MMP-2, is significantly higher compared with $Rln1^{+/+}$ mice (Table 1). Despite this increased level of MMP expression in the cervix of $Rln1^{-/-}$ mice, there is no clear histological evidence of collagen degradation in this tissue. Interestingly, when pregnant $Rln1^{-/-}$ mice are treated exogenously with a chronic infusion of human relaxin, there are significant decreases in cervical MMP-13 gene expression and no effects on MMP-2.[35] These findings are not dissimilar from earlier work in the pig cervix which described negative effects of relaxin treatment on tissue-associated MMP-2 and MMP-9 activity and no difference in gelatinase protein expression between control and relaxin treated animals.[43] Recent data in the rhesus monkey demonstrated that relaxin negatively regulates endometrial MMP-1 and MMP-3 protein expression in vivo.[44] These data are in contrast to the relaxin-induced increases in MMP-2 reported in nonreproductive tissues.[45-47] One problem with this data is that it only demonstrates MMP activity. But this gives no measure of the interaction between relaxin and MMP production, so the interpretation is limited to an association with MMP activation. The one exception is the work of Conrad and colleagues who clearly demonstrated the positive effects of relaxin treatment in male rats on MMP-2 gene and protein expression and MMP-2 activity in small renal arteries.[48,49]

The processes by which MMPs regulate the extracellular matrix are complex and multifactorial. One aspect is the role of tissue inhibitors of metalloproteinases (TIMPs), which directly regulate MMP activity. In the pig uterine cervix, relaxin enhanced expression of TIMP-1 and TIMP-2, whereas expression of both TIMPs in the vaginal cervix did not differ between control and relaxin-treated animals.[50] Relaxin treatment also increases TIMP-1 in the endometrium of the rhesus monkey.[44] The expression of TIMPs in reproductive tissues of $Rln1^{-/-}$ mice has not been assessed, but perhaps inhibitor activity of TIMPs is enhanced. In summary, relaxin's interactions with MMPs remain an area of controversy but in the in vivo data do not support a stimulatory effect of relaxin on MMP expression in reproductive tissues. It is, therefore, unlikely that relaxin acts via MMPs to reduce the density of collagen fibers in the cervix.

Structural changes in the cervix during pregnancy are not limited to the extracellular matrix. The proliferative activity of the cervical epithelium increases at two well-differentiated time points in the pregnant rat.[51,52] The first occurs in mid-gestation and the second close to delivery. The effect of this cell proliferation is to increase the height of the luminal epithelium. In the cervical stroma, cell proliferation is generally much lower compared with the epithelium, but there is a net increase in cell number because very few cells undergo apoptosis.[52] This cervical cell proliferation is attributed, in part, to a decrease in the rate of programmed cell death. The rate of apoptosis in both cervical compartments varies between stages of pregnancy. Lee et al[53] showed that the apoptotic index in the cervical epithelium was approximately 2% in early gestation in the rat, declined to approximately 0.5% in mid-late gestation and increased dramatically to 18% by the second day after delivery. Similarly, Ramos et al[52] reported that the apoptotic activity in the cervical epithelium never exceeded 1.8%, with the highest scores for programmed cell death on day 5 and the lowest indices between days 13 and 23. A dramatic increase in epithelial apoptosis was observed on the day after parturition, reaching values of 9%.[52] Stromal compartments had increased apoptotic indices postpartum, as seen in the epithelium. But the apoptotic rate in the cervical stroma was always lower than in the epithelium and was not observed in endothelial cells. These data strengthen the current hypothesis that apoptosis plays a major role in regulating cervical epithelial and stromal cell proliferation during pregnancy.

Relaxin plays an important function is stimulating cell proliferation and reducing apoptosis in the cervical epithelium and stroma during late pregnancy.[53,54] It promotes a marked increase in the accumulation of new epithelial and stromal cells.[54] One explanation of these data is the direct effect of relaxin on programmed cell death. Immunoneutralization of endogenous relaxin in late pregnant rats with MCA1 increased the rate of apoptosis in cervical cells.[55] The effect was greatest in late pregnancy when the rates of apoptosis in cervical epithelial cells and stromal cells were up to 10-fold higher in MCA1-treated rats compared with controls.[53]

Effects of Relaxin on the Vagina

Many of the actions of relaxin in the vagina are similar to those in the cervix. Early studies demonstrated that relaxin promoted growth of the vagina in pregnant rats, mice and pigs (reviewed in ref. 2). This was shown by the increase in wet and dry weights and vaginal collagen content.[56] In both MCA1-treated rats and $Rln1^{-/-}$ mice, the collagen fibers do not disperse and there is a marked lack of epithelial proliferation.[39,55] Administration of relaxin to $Rln1^{-/-}$ mice reverses this phenotype and in particular stimulates a dramatic increase in epithelial cell number. The recent work of Sherwood and colleagues has clearly demonstrated that the increase in vaginal epithelial cell number in the late pregnant rat involves an inhibition of apoptosis.[55] A physiological role for relaxin in the vagina has not been defined as such, but the relaxin-induced morphological changes are likely to facilitate the process of parturition.

Effects of Relaxin on the Uterus

Relaxin has been dismissed as an important player in uterine physiology largely because pups of $Rln1^{-/-}$ and $Lgr7^{-/-}$ mice are born alive, with no delay and within normal birth weights.[6,21] Furthermore, women who become pregnant through ovum donation have normal pregnancies

Table 2. A summary of the putative factors involved in the different mechanisms of relaxin action in the uterus associated with angiogenesis, implantation and growth

	Stromal Cells	Epithelial Cells	Decidual Cells	Uterus
VEGF	↑[82,83]	↑ & ↓[83]		↑[100]
Interleukin-11	↑[77]			
IGF-I, IGF-II				↑ secretion only[66]
IGFBPs	↔ but ↑ with MPA[66,76,78]		↑[29,78]	↑ secretion only[66]
E-cadherin		↑[69]		↑[69]
Connexins				↑[68]
Prolactin	↑[75,77,78]		↑[78]	
ER alpha				↓[44] & ↔[24,71]
ER beta				↔[44] & ↓[24,71]
PR				↓[44]

despite having no circulating relaxin.[57] However, it is important to recognize that experiments involving administration of exogenous relaxin have yielded some important functional data. These actions of relaxin in the endometrium, myometrium and placenta have been reviewed extensively,[2,8,58] so this section will focus on the proposed mechanisms of relaxin action, particularly in the endometrium (Table 2).

Uterine Growth

The growth effects of relaxin on the uterus have been well described in many species. Relaxin causes an increase in water content, protein, collagen and glycogen concentration in the uteri of estrogen primed, nonpregnant rats and an increase in uterine weight.[59-62] In addition, relaxin promotes uterine growth in the prepubertal[63,64] and neonate pig.[65] There is evidence to suggest that the growth effects of relaxin in prepubertal pigs are mediated by insulin-like growth factors (IGFs) and IGF-binding proteins (IGFBPs).[66,67] Uterine fluids collected from relaxin-treated gilts contained significantly higher amounts of IGF-I, IGF-II, IGFBP-2 and IGFBP-3 compared with controls.[66] However, relaxin administration did not alter IGF-I or-II gene expression in uterine tissue or systemic IGFs and IGFBPs. These data demonstrate a mechanism by which relaxin could contribute to uterine and conceptus growth in the early establishment of pregnancy. The work of Bagnell and colleagues in the prepubertal pig model also demonstrated relaxin-induced increases in gap junction proteins connexins and the glycoprotein E-cadherin, both thought to be important mediators of uterine growth and remodeling. Specifically, relaxin administration enhanced the expression of connexin-26, -32 and -43.[68] It was suggested by these authors that relaxin may mediate cell-cell communication between endothelial cells and the surrounding stroma and smooth muscle by increasing connexin protein expression. Relaxin-induced uterine growth in the prepubertal pig is also associated with a significant increase in epithelial cadherin (E-cadherin) protein and mRNA levels.[69] This calcium-dependent adhesion molecule is thought to mediate cell-to-cell recognition and maintain tissue integrity. The prepubertal pig is an interesting model because it lacks the local or systemic steroid hormones progesterone and estradiol. Therefore, it does not necessarily replicate the endocrine environment of pregnancy in many species. However, it has highlighted a number of novel mechanisms through which relaxin is capable of producing growth effects in the uterus.

In other animal models, relaxin's growth-promoting effects in the uterus are largely dependent on estrogen and progesterone. When administered with these steroids, relaxin stimulates growth by causing both cellular hyperplasia and hypertrophy.[70] More recent studies demonstrated that the uterotropic effects of relaxin are blocked by the specific estrogen receptor (ER)α antagonist ICI 182,780 in immature ovariectomized rats,[62] and that relaxin treatment decreases uterine ERβ

expression within 6 hours, but has no effect on ERα.[71] Treatment of Rln1$^{-/-}$ mice from day 12.5 gestation with a continuous infusion of recombinant human relaxin for 6 days has no effect on ERα gene expression but causes a significant down-regulation in ERβ expression and reverses the ERβ phenotype observed in Rln1$^{-/-}$ mice.[24] These data mirror the data of Pillai et al,[71] and support their idea that a down-regulation in ERβ expression by relaxin may be essential to allow for full activation and/or expression of ERα in the uterus. Several groups have reported that both ERβ1 and ERβ2 inhibit ERα-mediated transcriptional activity or signaling,[72,73] so a relaxin-induced down-regulation of ERβ may be a prerequisite for estrogen and other ER activators to stimulate their target tissues. However, this idea was recently challenged in ovariectomized rhesus monkeys that were primed with exogenous estradiol and progesterone in a manner that simulated a human menstrual cycle. Relaxin treatment significantly decreased uterine protein levels of ERα and both isoforms of the progesterone receptor, but had no effect on ERβ.[44] These authors concluded that relaxin may be responsible for the decline in endometrial expression of ERα and progesterone receptors that occurs during the late secretory phase of the human cycle. As with many responses to relaxin, there are large differences between species. It is possible that the variation in the response of ERs to relaxin treatment is due to the different experimental paradigms used, including stage of reproductive cycle and circulating steroid hormone levels. Furthermore, it is yet to be established how the peptide hormone activates steroid receptors in vivo.

Decidualization

Early studies using human endometrial stromal cells demonstrated that relaxin stimulates prolactin secretion after administration of relaxin.[74] Relaxin does not result in endometrial stromal cell growth, only increases prolactin production.[75] In order for relaxin to have an effect in stromal cells, a progestin (MPA) needs to be present. Relaxin also causes a transient stimulation of prolactin and IGFBP-1 mRNA within the endometrium. However, when relaxin was administered with MPA, higher mRNA levels were measured then if cells were treated with MPA alone or had MPA withdrawn.[76] More recent studies have further demonstrated relaxin's potential involvement in decidualization as treatment of human endometrial stromal cells with the hormone increases interleukin-11 mRNA expression and secretion via cAMP/protein kinase A pathways.[77] These authors proposed that relaxin acts in synergy with prostaglandin E2 to stimulate interleukin-11 production in the mid-late secretory phase of the cycle, before prolactin is detected and may therefore initiate endometrial cell differentiation. In other work, relaxin promotes induction of IGFBP-1 by binding to the cAMP regulatory element (CRE) in the IGFBP-1 promoter.[78] However, a progestin (MPA) needs to be administered together with relaxin,[78] or stromal cells need to be transfected with an LGR7 expression vector, in order to increase IGFBP-1 expression.[29] Tseng and colleagues also reported that relaxin increased the phosphorylation of CRE binding protein, indicating that relaxin activates the protein kinase A system. The complex nature of relaxin's interaction with the IGFBP-1 promoter is further demonstrated by studies using protein kinase A inhibitors. Relaxin-induced IGFBP-1 promoter activity was inhibited by the cAMP dependent protein kinase A inhibitor, H-89. Similarly, activation of prolactin by relaxin appears to be mediated through the region in the prolactin promoter containing multiple CCAAT/enhancer-binding proteins (C/EBP) binding sites.[79] Prolactin promoter activity was also inhibited by protein kinase A inhibitors. In summary, Tseng and colleagues proposed that relaxin acts via protein kinase A-dependent signaling pathways to activate two markers of decidualization, IGFBP-1 and prolactin.

Uterine Vascularization

New roles for relaxin as a vascular hormone within the uterus have also been established. Relaxin promotes endometrial and placental growth,[8,80] and may increase uterine blood flow in early pregnancy.[81] There are two explanations for these effects. The first is that relaxin causes vascular development or proliferation (angiogenesis), a view supported by in vitro cell culture studies using human endometrial stromal cells. The current hypothesis is that relaxin regulates uterine angiogenesis via vascular endothelial growth factor (VEGF). Relaxin upregulates VEGF gene expression and secretion from human endometrial stromal and glandular epithelial cells[7,82,83] by activating the VEGF promoter region.[83] This may occur via ERα, hypoxia inducible factor 1 alpha (HIF-1α) or SP1.[83,84] Increased vascularization in prostate xenograft tumors that over-express human relaxin is also associated with elevated VEGF gene expression.[85]

This work on angiogenesis was extended to an in vivo primate model, to demonstrate that relaxin treatment stimulated new blood vessel formation in the endometrium (Fig. 4 in ref. 44). These data strengthened much of the early work in ovariectomized rats and monkeys treated with porcine relaxin in combination with estrogen. It was only when animals were pretreated with estrogen that relaxin increased arteriole number per unit area and dilated blood vessels on the endometrial luminal surface.[86] It also caused a thickening of blood vessels and the proliferation of endothelial cells in arterioles and capillaries in the endometrium.[86-88] Relaxin increases vascularization in immature rats by enlarging the diameter of arteries and veins in the area between the circular and muscular sections of the uterus, thus providing increased blood flow.[61] This data was placed in context of human physiology in a phase II/III clinical trial for the treatment of scleroderma. Women receiving human relaxin reported heavier or irregular menstrual bleeding, indicating increased endometrial vascularization.[82,89]

Relaxin has also been implicated in the regulation of uterine blood flow. A direct effect of the peptide on uterine artery relaxation has been shown in mid-pregnant rats,[90] and it increases uterine artery blood flow in conscious, ovariectomized nonpregnant rats.[91] Furthermore, in vitro analysis of uterine artery vasodilation demonstrated that treatment with relaxin increased vessel diameter in response to elevated intraluminal pressure.[91] As discussed previously, relaxin binding sites are localized to blood vessels in the pig and human uterus,[16,25,92] and on blood vessels within the human amnion and placental villi.[25] Recent work has shown Lgr7 gene and protein expression in the aorta, mesenteric and small renal arteries of nonpregnant rats and mice.[93] Therefore, relaxin could be acting directly on Lgr7 in uterine arteries to mediate vasodilation and increase uterine blood flow to the placenta.

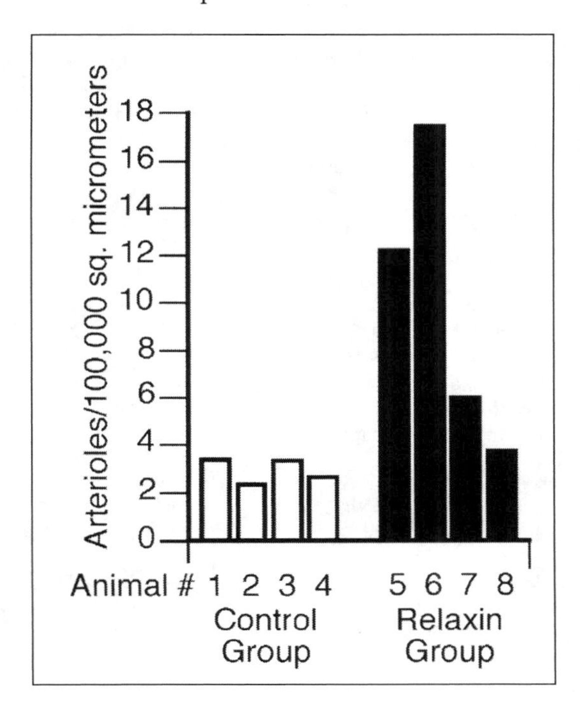

Figure 4. The effects of relaxin treatment on arteriole number in the endometrium of ovariectomized, steroid-primed rhesus monkeys. Reprinted with permission from L.T. Goldsmith et al. PNAS; 101:4685-4689. ©2004 The National Academy of Sciences (USA).

Placental Growth

In humans, relaxin is produced in low concentrations by intrauterine tissues such as the amnion, chorion, decidua, basal plate and placental trophoblast.[94] Therefore it is likely that relaxin acts as an autocrine or paracrine hormone to influence placental tissue growth.[95] Relaxin stimulates IGF-II to cause proliferation of human amniotic epithelial (WISH) cells in vitro.[80] Relaxin-treated WISH cells failed to proliferate when an antibody for IGF-II was added.[80] In addition, an in vivo component of this study identified that an increase in relaxin mRNA expression levels was correlated with a larger fetal membrane surface area and neonatal birth weight.[80] Small for gestational age infants have smaller placentas than controls, signifying that placental size is an indicator of fetal growth rate.[96] In conclusion, Millar et al suggested that relaxin could be an indicator of normal placental size and fetal growth rate. Of interest, increased risk of spontaneous abortion in horses is associated with placental insufficiency and low placental relaxin levels,[97,98] especially in mares with twin fetuses.[98,99] No study to date has examined placental growth or endometrial function in relaxin-deficient mice, nor have measurements been taken of fetal growth during pregnancy. Preliminary data from our laboratory indicate a lower conceptus weight in early gestation in $Rln1^{-/-}$ mice and a 7% reduction in fetal weight in $Rln1^{-/-}$ mice on day 18.5 gestation (1 day before expected births in mice). However, relaxin is not essential for implantation because fetuses develop to term and the average litter size is not different from wild-type mice.[6]

Summary

In order for pregnancy to be maintained until the appropriate time for parturition, many changes need to take place within the maternal reproductive tract. During early pregnancy, the endometrial stromal cells decidualize around the time of implantation and the placenta forms. In addition, uterine blood flow increases and the vascular bed proliferates to maintain a good supply of oxygen and nutrients to the fetus. The uterus also increases in size to accommodate the growing fetus and the myometrium remains quiescent to prevent premature contractions. At the end of pregnancy, the cervix softens and ripens to enable it to dilate during birth and the myometrium switches to a contractile apparatus. Although not common to all species, relaxin has been shown to play key roles in all these aspects of female reproductive physiology. In some species, a lack of relaxin can have serious implications for the maintenance of pregnancy and the birth of live young. This review has highlighted that relaxin treatment in a variety of pregnant animals has several stimulatory effects on growth factors, angiogenic factors and extracellular matrix components that could be perceived as beneficial for establishing and maintaining a successful pregnancy as well as facilitating the process of labor.

Acknowledgments

We are grateful to the Australian Research Council and the Department of Zoology (University of Melbourne) for recent funding to support the work in our laboratory. Lenka Vodstrcil currently holds an NH&MRC Dora Lush postgraduate scholarship. The human relaxin used in our in vivo mouse studies is a gift from BAS Medical. We thank Jonathan McGuane, Helen Gehring, Kirk Conrad, Mary Wlodek and Elaine Unemori for their invaluable contributions to our research over the last few years.

References

1. Hisaw FL. Experimental relaxation of the pubic ligament of the guinea pig. Proc Soc Exp Biol Med 1926; 23:661-663.
2. Sherwood OD. Relaxin's physiological roles and other diverse actions. Endocr Rev 2004; 25(2):205-234.
3. Guico-Lamm ML, Voss EWJ, Sherwood OD. Monoclonal antibodies specific for rat relaxin. II. Passive immunization with monoclonal antibodies throughout the second half of pregnancy disrupts birth in intact rats. Endocrinology 1988; 123(5):2479-2485.

4. Hwang JJ, Sherwood OD. Monoclonal antibodies specific for rat relaxin. III. Passive immunization with monoclonal antibodies throughout the second half of pregnancy reduces cervical growth and extensibility in intact rats. Endocrinology 1988; 123(5):2486-2490.
5. Kuenzi MJ, Sherwood OD. Monoclonal antibodies specific for rat relaxin. VII. Passive immunization with monoclonal antibodies throughout the second half of pregnancy prevents development of normal mammary nipples morphology and function in rats. Endocrinology 1992; 131(4):1841-1847.
6. Zhao L, Roche PJ, Gunnerson JM et al. Mice without a functional relaxin gene are unable to deliver milk to their pups. Endocrinology 1999; 140(1):445-453.
7. Weiss G, Palejwala S, Tseng L et al. Synthesis and function of relaxin in human endometrium. In: Tregear GW, Ivell R, Bathgate RA, Wade JD, eds. Relaxin 2000. Dordrecht, The Netherlands: Kluwer Academic Publishers, 2001:41-45.
8. Hayes ES. Biology of primate relaxin: a paracrine signal in early pregnancy? Reprod Biol Endocrinol 2004; 2:36.
9. Hsu SY, Nakabayashi K, Nishi S et al. Activation of orphan receptors by the hormone relaxin. Science 2002; 295:671-674.
10. Hsu SYT. New insights into the evolution of the relaxin-LGR signalling system. Trends Endocrin Met 2003; 14(7):303-309.
11. Scott DJ, Layfield S, Riesewijk A et al. Identification and characterization of the mouse and rat relaxin receptors as the novel orthologues of human leucine-rich repeat-containing G-protein-coupled receptor 7. Clin Exp Pharmacol Physiol 2004; 31(11):828-832.
12. Bathgate RA, Ivell R, Sanborn BM et al. International Union of Pharmacology LVII: recommendations for the nomenclature of receptors for relaxin family peptides. Pharmacol Rev 2006; 58(1):7-31.
13. Dessauer CW, Nguyen BT. Relaxin stimulates multiple signaling pathways: activation of cAMP, PI3K and PKCzeta in THP-1 cells. Ann NY Acad Sci 2005; 1041:272-279.
14. Nguyen BT, Dessauer CW. Relaxin stimulates protein kinase C zeta translocation: requirement for cyclic adenosine 3',5'-monophosphate production. Mol Endocrinol 2005; 19(4):1012-1023.
15. Kuenzi MJ, Connolly BA, Sherwood OD. Relaxin acts directly on rat mammary nipples to stimulate their growth. Endocrinology 1995; 136(7):2943-2947.
16. Min G, Sherwood OD. Identification of specific relaxin-binding cells in the cervix, mammary glands, nipples, small intestine and skin of pregnant pigs. Biol Reprod 1996; 55:1243-1252.
17. Mercado-Simmen RC, Bryant-Greenwood GD, Greenwood FC. Characterization of the binding of 125I-relaxin to rat uterus. J Biol Chem 1980; 225:3617-3623.
18. Mercado-Simmen RC, Bryant-Greenwood GD, Greenwood FC. Relaxin receptor in the rat myometrium: regulation by estrogen and relaxin. Endocrinology 1982; 110(1):220-226.
19. Tan YY, Wade JD, Tregear GW et al. Quantitative autoradiographic studies of relaxin binding in rat atria, uterus and cerebral cortex: characterization and effects of oestrogen treatment. Br J Pharmacol 1999; 127(1):91-98.
20. Hsu SY, Kudo M, Chen T et al. The three subfamilies of leucine-rich repeat-containing G protein-coupled receptors (LGR): identification of LGR6 and LGR7 and the signalling mechanism for LGR7. Mol Endocrinol 2000; 14(8):1257-1271.
21. Krajnc-Franken MAM, van Disseldorp JM, Koenders JE et al. Impaired nipple development and parturition in LGR7 knockout mice. Mol Cell Biol 2004; 24(2):687-696.
22. Feng S, Bogatcheva NV, Kamat AA et al. Endocrine effects of relaxin overexpression in mice. Endocrinology 2006; 147(1):407-414.
23. Siebel AL, Gehring HM, Reytomas IGT et al. Inhibition of oxytocin receptor and estrogen receptor-α, but not relaxin receptors (LGR7), in the myometrium of late pregnant relaxin gene knockout mice. Endocrinology 2003; 144(10):4272-4275.
24. Siebel AL, Gehring HM, Vodstrcil L et al. Oxytocin and estrogen receptor expression in the myometrium of pregnant relaxin-deficient (Rlx-/-) mice. Ann NY Acad Sci 2005; 1041:104-109.
25. Kohsaka T, Min G, Lukas G et al. Identification of specific relaxin-binding cells in the human female. Biol Reprod 1998; 59:991-999.
26. Ivell R, Balvers M, Pohnke Y et al. Immunoexpression of the relaxin receptor LGR7 in breast and uterine tissues of humans and primates. Reprod Biol Endocrinol 2003; 1:114-126.
27. Luna JJ, Riesewijk A, Horcajadas JA et al. Gene expression pattern and immunoreactive protein localization of LGR7 receptor in human endometrium throughout the menstrual cycle. Mol Hum Reprod 2004; 10(2):85-90.
28. Bond CP, Parry LJ, Samuel CS et al. Increased expression of the relaxin receptor (LGR7) in human endometrium during the secretory phase of the menstrual cycle. J Clin Endocr Metab 2004; 89(7):3477-3485.

29. Mazella J, Tang M, Tseng L. Disparate effects of relaxin and TGF beta1: relaxin increases, but TGF beta1 inhibits, the relaxin receptor and the production of IGFBP-1 in human endometrial stromal/decidual cells. Hum Reprod 2004; 19(7):1513-1518.
30. Einspanier A, Muller D, Lubberstedt J et al. Characterization of relaxin binding in the uterus of the marmoset monkey. Mol Hum Reprod 2001; 7(10):963-970.
31. Lowndes K, Amano A, Yamamoto SY et al. The human relaxin receptor (LGR7): expression in the fetal membranes and placenta. Placenta 2006; 27(6-7):610-618.
32. Scott DJ, Layfield S, Yan Y et al. Characterization of novel splice variants of LGR7 and LGR8 reveals that receptor signaling is mediated by their unique low density lipoprotein class A modules. J Biol Chem 2006; 281(46):34942-34954.
33. Sherwood OD. Multiple physiological effects of relaxin during pregnancy. In: Bazer FW, ed. The Endocrinology of Pregnancy. Totowa, New Jersey: Human Press Inc.; 1998:431-460.
34. Sherwood OD, Zhao S. Effects of relaxin on the cervix and vagina in the rat. In: Tregear GW, Ivell R, Bathgate RA, Wade JD, eds. Relaxin 2000. Dordrecht, The Netherlands: Kluwer Acadmic Publishers; 2001:53-58.
35. Parry LJ, McGuane JT, Gehring HM et al. Mechanisms of relaxin action in the reproductive tract: studies in the relaxin-deficient (Rlx-/-) mouse. Ann N Y Acad Sci 2005; 1041:91-103.
36. Downing SJ, Sherwood OD. The physiological role of relaxin in the pregnant rat. IV. The influence of relaxin on cervical collagen and glycosaminoglycans. Endocrinology 1986; 118(2):471-479.
37. O'Day MB, Winn RJ, Easter RA et al. Hormonal control of the cervix in pregnant gilts. II. Relaxin promotes changes in the physical properties of the cervix in ovariectomized hormone-treated pregnant gilts. Endocrinology 1989; 125(6):3004-3010.
38. O'Day-Bowman MB, Winn RJ, Dziuk PJ et al. Hormonal control of the cervix in pregnant gilts. III. Relaxin's influence on cervical biochemical properties in ovariectomized hormone-treated pregnant gilts. Endocrinology 1991; 129(4):1967-1976.
39. Zhao L, Samuel CS, Tregear GW et al. Collagen studies in late pregnant relaxin null mice. Biol Reprod 2000; 63:697-703.
40. Mushayandebvu TI, Rajabi MR. Relaxin stimulates interstitial collagenase activity in cultured uterine cervical cells from nonpregnant and pregnant but not immature guinea pigs; estradiol-17 beta restores relaxin's effect in immature cervical cells. Biol Reprod 1995; 53(5):1030-1037.
41. Palejwala S, Stein DE, Weiss G et al. Relaxin positively regulates matrix metalloproteinase expression in human lower uterine segment fibroblasts using a tyrosine kinase signaling pathway. Endocrinology 2001; 142(8):3405-3413.
42. Hwang JJ, Macinga D, Rorke EA. Relaxin modulates human cervical stromal cell activity. J Clin Endocrinol Metab 1996; 81(9):3379-3384.
43. Lenhart JA, Ryan PL, Ohleth KM et al. Relaxin increases secretion of matrix metalloproteinase-2 and matrix metalloproteinase-9 during uterine and cervical growth and remodeling in the pig. Endocrinology 2001; 142(9):3941-3949.
44. Goldsmith LT, Weiss G, Palejwala S et al. Relaxin regulation of endometrial structure and function in the rhesus monkey. Proc Natl Acad Sci USA 2004; 101(13):4685-4689.
45. Mookerjee I, Unemori EN, Du XJ et al. Relaxin modulates fibroblast function, collagen production and matrix metalloproteinase-2 expression by cardiac fibroblasts. Ann N Y Acad Sci 2005; 1041:190-193.
46. Samuel CS. Relaxin: antifibrotic properties and effects in models of disease. Clin Med Res 2005; 3(4):241-249.
47. Silvertown JD, Walia JS, Summerlee AJ et al. Functional expression of mouse relaxin and mouse relaxin-3 in the lung from an Ebola virus glycoprotein-pseudotyped lentivirus via tracheal delivery. Endocrinology 2006; 147(8):3797-3808.
48. Jeyabalan A, Novak J, Danielson LA et al. Essential role for vascular gelatinase activity in relaxin-induced renal vasodilation, hyperfiltration and reduced myogenic reactivity of small arteries. Circulation Research 2003; 93:1249-1257.
49. Jeyabalan A, Kerchner LJ, Fisher MC et al. Matrix metalloproteinase-2 activity, protein, mRNA and tissue inhibitors in small arteries from pregnant and relaxin-treated nonpregnant rats. J Appl Physiol 2006; 100(6):1955-1963.
50. Lenhart JA, Ryan PL, Ohleth KM et al. Relaxin increases secretion of tissue inhibitor of matrix metalloproteinase-1 and -2 during uterine and cervical growth and remodeling in the pig. Endocrinology 2002; 143(1):91-98.
51. Burger LL, Sherwood OD. Evidence that cellular proliferation contributes to relaxin-induced growth of both the vagina and the cervix in the pregnant rat. Endocrinology 1995; 136(11):4820-4826.
52. Ramos JG, Varayoud J, Bosquiazzo VL et al. Cellular turnover in the rat uterine cervix and its relationship to estrogen and progesterone receptor dynamics. Biol Reprod 2002; 67:735-742.

53. Lee HY, Zhao S, Fields PA et al. The extent to which relaxin promotes proliferation and inhibits apoptosis of cervical epithelial and stromal cells is greatest during late pregnancy in rats. Endocrinology 2005; 146(1):511-518.
54. Burger LL, Sherwood OD. Relaxin increases the accumulation of new epithelial and stromal cells in the rat cervix during the second half of pregnancy. Endocrinology 1998; 139(9):3984-3995.
55. Zhao S, Fields PA, Sherwood OD. Evidence that relaxin inhibits apoptosis in the cervix and the vagina during the second half of pregnancy in the rat. Endocrinology 2001; 142(6):2221-2229.
56. Zhao S, Kuenzi M, Sherwood O. Monoclonal antibodies specific for rat relaxin. IX. Evidence that endogenous relaxin promotes growth of the vagina during the second half of pregnancy in rats. Endocrinology 1996; 137(2):425-430.
57. Johnson MR, Abdalla H, Allman AC et al. Relaxin levels in ovum donation pregnancies. Fertil Steril 1991; 56(1):59-61.
58. Relaxin and related peptides. New York: New York Academy of Sciences; 2005.
59. Steinetz BG, Beach VL, Blye RP et al. Changes in the composition of the rat uterus following a single injection of relaxin. Endocrinology 1957; 61:287-292.
60. Brennan DM, Zarrow MX. Water and electrolyte content of the uterus of the intact and adrenalectomized rat treated with relaxin and various steroid hormones. Endocrinology 1959; 64:907-913.
61. Vasilenko P, Mead JP, Weidmann JE. Uterine growth-promoting effects of relaxin: a morphometric and histological analysis. Biol Reprod 1986; 35:987-995.
62. Pillai SB, Rockwell C, Sherwood OD et al. Relaxin stimulates uterine edema via activation of estrogen receptors: blockade of its effects using ICI 182, 780, a specific estrogen receptor antagonist. Endocrinology 1999; 140(5):2426-2429.
63. Hall JA, Cantley TC, Day BN et al. Uterotropic actions of relaxin in prepubertal gilts. Biol Reprod 1990; 42(5-6):769-774.
64. Bagnell CA, Zhang Q, Downey B et al. Sources and biological actions of relaxin in pigs. J Reprod Fertil 1993; 48:127-138.
65. Bagnell CA, Yan W, Wiley AA et al. Effects of relaxin on neonatal porcine uterine growth and development. Ann N Y Acad Sci 2005; 1041:248-255.
66. Ohleth KM, Lenhart JA, Ryan PL et al. Relaxin increases insulin-like growth factors (IGFs) and IGF-binding proteins of the pig uterus in vivo. Endocrinology 1997; 138(9):3652-3658.
67. Ohleth KM, Zhang Q, Lenhart JA et al. Trophic effects of relaxin on reproductive tissus: role of the IGF system. Steroids 1999; 64(9):634-639.
68. Lenhart JA, Ryan PL, Ohleth KM et al. Expression of connexin-26, -32 and -43 gap junction proteins in the porcine cervix and uterus during pregnancy and relaxin-induced growth. Biol Reprod 1999; 61:1452-1459.
69. Ryan PL, Baum DL, Lenhart JA et al. Expression of uterine and cervical epithelial cadherin during relaxin-induced growth in pigs. Reproduction 2001; 122(6):929-937.
70. Brody S, Wiqvist N. Ovarian hormones and uterine growth: effects of estradiol, progesterone and relaxin on cell growth and cell division in the rat uterus. Endocrinology 1961; 68:971-977.
71. Pillai SB, Jones JM, Koos RD. Treatment of rats with 17 beta-estradiol or relaxin rapidly inhibits uterine estrogen receptor beta1 and beta2 messenger ribonucleic acid levels. Biol Reprod 2002; 67:1919-1926.
72. Hall JM, McDonnell DP. The estrogen receptor beta-isoform (ERbeta) of the human estrogen receptor modulates ERalpha transcriptional activity and is a key regulator of the cellular response to estrogens and antiestrogens. Endocrinology 1999; 140:5566-5578.
73. Lindberg MK, Moverare S, Skrtic S et al. Estrogen receptor (ER)-beta reduces ERalpha-regulated gene transcription, supporting a "ying yang" relationship between ERalpha and ERbeta in mice. Mol Endocrinol 2003; 17:203-208.
74. Huang JR, Tseng L, Bischof P et al. Regulation of prolactin production by progestin, estrogen and relaxin in human endometrial stromal cells. Endocrinology 1987; 121(6):2011-2017.
75. Zhu HH, Huang JR, Mazella J et al. Differential effects of progestin and relaxin on the synthesis and secretion of immunoreactive prolactin in long term culture of human endometrial stromal cells. J Clin Endocrinol Metab 1990; 71(4):889-899.
76. Tseng L, Gao JG, Chen R et al. Effect of progestin, antiprogestin and relaxin on the accumulation of prolactin and insulin-like growth factor-binding protein-1 messenger ribonucleic acid in human endometrial stromal cells. Biol Reprod 1992; 47(3):441-450.
77. Dimitriadis E, Stoikos C, Baca M et al. Relaxin and prostaglandin E(2) regulate interleukin 11 during human endometrial stromal cell decidualization. J Clin Endocrinol Metab 2005; 90(6):3458-3465.
78. Tang M, Mazella J, Zhu HH et al. Ligand activated relaxin receptor increases the transcription of IGFBP-1 and prolactin in human decidual and endometrial stromal cells. Mol Hum Reprod 2005; 11(4):237-243.

79. Pohnke Y, Kempf R, Gellersen B. CCAAT/enhancer-binding proteins are mediators in the protein kinase A-dependent activation of the decidual prolactin promoter. J Biol Chem 1999; 274(35):24808-24818.
80. Millar LK, Reiny R, Yamamoto SY et al. Relaxin causes proliferation of human amniotic epithelium by stimulation of insulin-like growth factor-II. Am J Obstet Gynecol 2003; 188(1):234-241.
81. Jauniaux E, Johnson MR, Jurkovic D et al. The role of relaxin in the development of the uteroplacental circulation in early pregnancy. Obstet Gynecol 1994; 84(3):339-342.
82. Unemori EN, Erikson ME, Rocco SE et al. Relaxin stimulates expression of vascular endothelial growth factor in normal human endometrial cells in vitro and is associated with menometrorrhagia in women. Hum Reprod 1999; 14(3):800-806.
83. Palejwala S, Tseng L, Wojtczuk A et al. Relaxin gene and protein expression and its regulation of procollagenase and vascular endothelial growth factor in human endometrial cells. Biol Reprod 2002; 66:1743-1748.
84. Koos RD, Kazi AA, Roberson MS, Jones JM. New insight into the transcriptional regulation of vascular endothelial growth factor expression in the endometrium by estrogen and relaxin. Ann N Y Acad Sci 2005; 1041:233-247.
85. Silvertown JD, Ng J, Sato T et al. H2 relaxin overexpression increases in vivo prostate xenograft tumor growth and angiogenesis. Int J Cancer 2006; 118(1):62-73.
86. Hisaw FL, Hisaw FL, Jr., Dawson AB. Effects of relaxin on the endothelium of endometrial blood vessels in monkeys (Macaca mulatta). Endocrinology 1967; 81(2):375-385.
87. Dallenbach-Hellweg G, Battista JV, Dallenbach FD. Immunohistological and histochemical localization of relaxin in the metrial gland of the pregnant rat. Am J Anat 1965; 117(3):433-450.
88. Einspanier A. Relaxin is an important factor for uterine differentiation and implantation in the marmoset monkey. In: Tregear GW, Ivell R, Bathgate RA, Wade JD, eds. Relaxin 2000. Dordrecht, The Netherlands: Kluwer Academic Publishers; 2001:73-82.
89. Seibold JR, Korn JH, Simms R et al. Recombinant human relaxin in the treatment of scleroderma. A randomized, double-blind, placebo-controlled trial. Ann Intern Med 2000; 132(11):871-879.
90. Longo M, Jain V, Vedernikov YP et al. Effects of recombinant human relaxin on pregnant rat uterine artery and myometrium in vitro. Am J Obstet Gynecol 2003; 188(6):1468-1476.
91. Novak J. Relaxin increases uterine blood flow in concious nonpregnant rats and decreases myogenic reactivity in isolated uterine arteries. FASEB. 2002; 16:A824.
92. Min G, Hartzog MG, Jennings RL et al. Evidence that endogenous relaxin promotes growth of the vagina and uterus during pregnancy in gilts. Endocrinology 1997; 138(2):560-565.
93. Conrad KP, Parry LJ, Shroff SG et al. Evidence that relaxin is a vascular-derived, locally-acting relaxing and compliance factor. J Soc Gynecol Invest 2005; 12:Suppl: 688A.
94. Sakbun V, Ali SM, Greenwood FC et al. Human relaxin in the amnion, chorion, decidua parietalis, basal plate and placental trophoblast by immunocytochemistry and northern analysis. J Clin Endocrinol Metab 1990; 70(2):508-514.
95. Liu HC, Mele CA, Noyes N et al. Endometrial secretory proteins enhance early embryo development. J Assist Reprod Genet 1994; 11(4):217-224.
96. Heinonen S, Taipale P, Saarikoski S. Weights of placentae from small-for-gestational age infants revisited. Placenta 2001; 22:399-404.
97. Stewart DR, Addiego LA, Pascoe DR et al. Breed differences in circulating equine relaxin. Biol Reprod 1992; 46(4):648-652.
98. Ryan PL, Vaala W, Bagnell CA. Evidence that equine relaxin is a good indicator of placental insufficiency in the mare. Paper presented at: Proceedings 44th Annual Convention of American Association of Equine Practitioners, 1998.
99. Ryan P, Vaala W, Bennett-Wimbush K et al. Relaxin and placental dysfunction in the horse. In: Tregear GW, Ivell R, Bathgate RA, Wade JD, eds. Relaxin 2000. Dordrecht, The Netherlands: Kluwer Academic Publishers; 2001:91-99.
100. Pillai SB, Sherwood OD, Koos RD. Relaxin up-regulates rat uterine vascular endothelial growth factor mRNA in vivo. In: Tregear GW, Ivell R, Bathgate RA, Wade JD, eds. Relaxin 2000. Dordrecht, The Netherlands: Kluwer Academic Publishers; 2001:109-113.

Chapter 5

Relaxin and Related Peptides in Male Reproduction

Alexander I. Agoulnik*

Abstract

The relaxin hormone is renowned for its function in pregnancy, parturition and other aspects of female reproduction. At the same time, the role of relaxin in male reproduction is still debated. Relaxin is prominently expressed in prostate and its receptors are found in several male reproductive organs; however, the data indicative of its contribution to differentiation and functioning of prostate or testis are contradictory. Prostate relaxin is a main source of this peptide in the seminal plasma. The relaxin effects on sperm motility and fertilization have been reported. The expression of other relaxin related peptides, such as INSL5 and INSL6 was described in testis; yet, currently there are no experimental data to pinpoint their biological functions. The other member of relaxin peptide family, insulin-like 3 peptide (INSL3), is a major player in male development. The INSL3 peptide is expressed in testicular fetal and adult Leydig cells and is directly responsible for the process of abdominal testicular descent (migration of the testes towards the scrotum during male development). Genetic targeting of the *Insl3* gene or INSL3 GPCR receptor *Lgr8/Rxfp2* causes high intra-abdominal cryptorchidism due to a differentiation failure of testicular ligaments, the gubernacula. Several mutations of these two genes rendering nonfunctional proteins have been described in human patients with testicular maldescent. Thus, in this chapter we review the data related to the expression and function of relaxin and related peptides in male reproduction.

Relaxin in Male Reproduction

Relaxin and Relaxin Receptor Expression in Male Reproductive Organs

The data accumulated to date strongly indicate that the main site of relaxin expression in mammals is the prostate and in some species the seminal vesicles, with a subsequent release of the relaxin peptide into the seminal fluid.[1,2] The immunoreactive relaxin or relaxin-like activity was identified in the seminal plasma of human, guinea pig, boars, baboons and other species.[1,3,4] Relaxin was detected in ejaculate of men with the congenital absence of testis and vas deference, but with, at least, partially functional prostate, proving evidence that prostate was the source of seminal relaxin.[1,4] Similarly, no decrease in relaxin serum concentration was detected in boars after castration, suggesting non-testicular source of the hormone in males.[5] Within the human prostate relaxin expression was detected exclusively in the glandular epithelium (Fig. 1).[6,7]

Both human *RLN1* and *RLN2* genes are expressed in the prostate at the mRNA level.[8-10] The analysis of the promoter activity of two relaxin genes in the prostate adenocarcinoma LNCaP cell line yielded rather contradictory results. Gunnersen et al showed that the promoter of

*Corresponding Author: Alexander I. Agoulnik—Department of Obstetrics and Gynecology, Baylor College of Medicine, Houston, Texas, 77030, U.S.A. Email: agoulnik@bcm.edu

Relaxin and Related Peptides, edited by Alexander I. Agoulnik. ©2007 Landes Bioscience and Springer Science+Business Media.

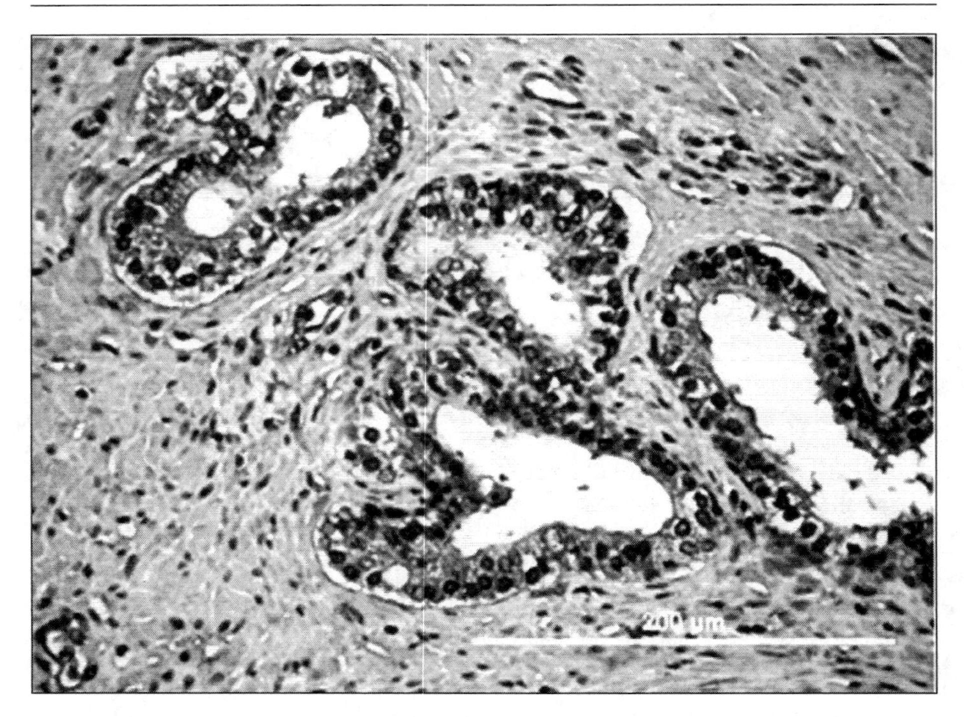

Figure 1. Expression of relaxin peptide in the glandular epithelium of human prostate. Immunohistochemistry with specific antibody to human relaxin 2 peptide shows the localization of relaxin in epithelial cells.

RLN2 was more active than that of *RLN1* in LNCaP cells.[11] On the other hand, Dr. Gillian D. Bryant-Greenwood and her colleagues showed that the basal level of *RLN2* mRNA was 10-fold lower than that of *RLN1*.[12] The latter group also established the differences in the stability of mRNA messages for two relaxin genes, as *RLN2* half-life was sevenfold shorter than that of *RLN1* mRNA.[12] The authors described the non-canonical poly (A) + signal in *RLN2* 3'-UTR that may also influence the mRNA level. The earlier reports indicated that only RLN2 peptide was present in the prostatic fluid.[13] Human seminal relaxin is a product of the same gene as human luteal relaxin, suggesting that *RLN1* may be a transcribed pseudogene.[14] Recently, the specific antibodies to the most diverged decapeptides based on the region of greatest difference between the two relaxins at the N-termini of the A-chains have been produced.[15] An expression of both relaxin isoforms was detected in the decidual cells, showing that RLN1 can be expressed at the protein level.[15] Further analysis with specific RLN1-reagents is needed to examine whether RLN1 is translated in the prostate.

Rather controversial results were obtained concerning relaxin expression in rodent prostate. The anti-rat and anti-porcin relaxin sera did not detect immunoreactivity neither in rat or mouse seminal plasma, testis, or in prostate, with parallel positive control staining in the ovary.[16] Nevertheless, the presence of relaxin mRNA was detected by in situ hybridization and by RT-PCR in rat and mouse prostate.[17,18] It should be noted, that the level of relaxin mRNA expression in rodent prostate is rather low in comparison with female reproductive organs (Parry, Agoulnik, Klonisch, unpublished data) and therefore the sensitivity of immunodetection might be limited.

Thompson et al showed that relaxin expression in prostate adenocarcinoma cell line LNCaP was downregulated by androgen treatment in vitro.[19] Castration of nude mice with transplanted LNCaP cells lead to an increase of relaxin expression. The authors showed also that the androgen-ablation therapy led to an increase of relaxin gene expression as was revealed by the immunostaining of prostate cancer tissue microarrays.[19] In summary, the results suggest the possible regulation of relaxin expression by androgens. The regulatory effect of other steroid hormones (progesterone and glicocorticoids) on relaxin promoter activity was also demonstrated.[20]

Immunoreactivity to relaxin antibodies and relaxin bioactivity was detected in testis from dogfish shark, armadillos and rooster. In mammals the expression of relaxin in testis remains uncertain. The early report indicated on association of relaxin immunoreactivity with interstitial and Sertoli cells of the boar testis,[21] however, these results were not confirmed in later studies.[22,23] At RNA level the mouse *Rln1* transcripts are detectable in adult testis and epididymis,[18] however, it remains to be shown whether the relaxin RNA is translated in these organs. In tamar wallaby testis, no relaxin expression was detected even by RT-PCR and subsequent Southern blot analysis of the resultant PCR fragments.[24] Thus, even if present, the relaxin expression in testis is rather low. Interestingly, in frogs the relaxin homolog is expressed in Leydig cells and is regulated by androgens.[25]

Relaxin receptor LGR7 expression was readily detected in numerous male reproductive organs at RNA level. The Northern blot analysis revealed the presence of a single 4.4kb band in adult human testis,[26] although several alternative splice variants of the gene with undefined functions have been described.[27-29] The effect of relaxin peptide on testicular somatic cells have not been analyzed in details, although, the specific relaxin-binding sites were determined in testicular Leydig cells of immature and mature boars.[30] We and others have shown that the functional relaxin receptor is expressed in normal human prostate;[19,31,32] its expression is maintained in benign prostate hyperplasia and in prostate cancer (Feng, Ittmann, Agoulnik, unpublished data). Stimulation with relaxin strongly increases migration and invasiveness, cell adhesion and survival of LNCaP and PC-3 prostate adenocarcinoma cell lines. We have shown that similar to breast cancer cells, the effect of relaxin on prostate cancer cell migration is MMP-dependent. The suppression of *LGR7* or *RLN2* RNA expression using short interference RNAs decreases invasive phenotype of the prostate cancer cells, their survival and increases cancer cell apoptosis in vitro. The prostate relaxin-overexpressing tumors transplanted into the nude mice exhibited greater tumor volumes compared to control tumors.[31] This was accompanied with an advanced angiogenic phenotype with greater intratumoral vascularization and increased VEGF expression. Such clear biological response of the prostate cancer cells to relaxin hormone strongly indicates a presence of functional relaxin receptors in these cells.

Deletion of Mouse Rln1 and Lgr7 and Development of Male Reproductive Organs

Mutant mice with genetic ablation of relaxin or Lgr7 receptor, as well as mice with transgenic overexpression of relaxin hormone have been generated.[33-35] Samuel et al reported that mice with relaxin gene deletion had significant abnormalities of male reproductive organs.[18] Mutant mice showed retarded growth and marked deficiencies in the reproductive tract with male reproductive organ weight (including the testis, epididymis, prostate and seminal vesicle) significantly smaller than those of wild-type male mice. This correlated with decrease in fertility, histologically detected decreased sperm maturation in testis, increased collagen content in the extracellular matrix of the testis and prostate, an increase in the rate of cell apoptosis and decreased epithelial proliferation in the prostate.[18] It should be noted, that in this study the authors did not specified which part of the mouse prostate was analyzed in the mutant and control animals.

Surprisingly, the effects of relaxin deficiency were not confirmed in the *Lgr7-/-*mutants. Detailed examination of reproductive organs at different stages of development, fertility rates, or histological analysis of testis development in Lgr7 mutants conducted in our laboratory failed to reveal male reproductive phenotype.[34] No effects of male fertility or reproductive characteristics were described in mice with transgenic overexpression of relaxin.[36] Similarly, in second independently generated *Lgr7*-deficient mutant, no deleterious effects on prostate development were described.[35] The authors described some transient effect on male fertility only in first two generations of the mutants; however, the decreased fertility was not detected in subsequent generations.[35]

What is the reason for the contradicting results obtained with *Rln1-/-* and *Lgr7-/-* mice? Several potential explanations can be suggested. First, the question arises whether there are additional receptors for relaxin in mice besides Lgr7. If relaxin signals through Lgr7-independent pathway, this may explain the difference in the phenotype of relaxin or Lgr7-deficient mice. To investigate this option we used mice with transgenic overexpression of relaxin. In these mice transgenic relaxin is expressed under insulin 2 promoter, causing increased mammary nipple development in females.[36] The mutant phenotype associated with the relaxin overproduction in females was completely abrogated on Lgr7-deficient background, indicating that Lgr7 is the only cognate receptor for relaxin in mice. Contrary to the in vitro effects of human and porcine relaxin, we showed that mouse relaxin did not signal through Insl3 receptor Lgr8; no prostate abnormalities were detected in mice with deletion of both Lgr7 and Lgr8.[34] The second possibility is that some subtle differences in genetic background between relaxin and Lgr7 mutant strains may be involved facilitating male reproductive effects in the first one. To address this hypothesis we transferred the mutant Lgr7 allele on C57BL/6 inbred background by backcrossing the mutant mice with C57BL/6 animals. Indeed, the female-specific phenotype of Lgr7-deficiency, such as abnormal parturition, was more pronounced on C57BL/6 inbred background. However, again, we were not able to detect any male-specific phenotype in the mutant animals from this strain. Finally, it is possible that the transient effects in the first generations of transgenic mutant mice were in fact not due to the relaxin-deficiency, but were the result of genetic rearrangements in other, perhaps even linked to the *Rln1* gene, parts of the mouse genome. These additional mutations might be lost from the genome after establishment of the mutant strains. In support of the latter explanation, it was reported recently, that careful re-examination of different prostate parts and fertility in relaxin-deficient mutants did not revealed any differences in comparison to the wild-type littermates from the same strain (Ganesan, Parry, unpublished data). The further examination of relaxin deficiency on male reproduction in the mutant mice is certainly required.

Relaxin Effects on Sperm

The presence of relaxin in seminal fluid suggests the role of this peptide in sperm functions.[1] The recombinant human relaxin binds with high affinity to human sperm indicating the presence of functional receptors on cell surface.[37] As mentioned above the relaxin is readily detectable in human and boar semen samples.[38-40] Recently, the immunoreactive relaxin (using anti-porcin antibody) was also detected in bulls, rams and he-goats.[41] The studies aimed to established the correlations between sperm quality and the relaxin concentration in seminal plasma have been conducted. The results of such analysis varied and appear to be species-specific. Whereas in boars the correlation between relaxin seminal concentration and sperm motility characteristics have been demonstrated,[1,40] no link between the two were obtained for human semen.[38,39,42] In bulls the relaxin level in semen was significantly correlated with the percentage of spermatozoa showing only the most intensive motility.[41]

To study the effects of relaxin in vitro two main approaches have been used: a) suppression of relaxin signaling in sperm using specific anti-relaxin antibody; b) direct treatment of sperm with relaxin peptide. The addition of relaxin antiserum significantly inhibited motility of the human and boar sperm.[1,43] From the other hand, the relaxin treatment of sperm had variable effects. Some studies did not produce any evidence of relaxin effect on motility of human,[38,42] bull,[44] or rabbit[45] sperm. On the contrary, the other groups reported a statistically significant effects of relaxin on human and boar sperm motility.[46-50] The stimulatory effect of relaxin was shown on motility of washed sperm, sperm aged for several hours and cryopreserved sperm.[46-48,51] It was reported that treatment with porcine relaxin increased the penetration of human sperm into cervical mucus,[1,52] the rates of in vitro fertilization of mouse oocytes,[53] as well as the penetration capacity of human spermatozoa.[51] Similar effects of relaxin on in vitro fertilization ability of boar spermatozoa, acrosome reaction and utilization of glucose were recently reported.[50,54] However, Chan and Tang (1984) did not find an effect of relaxin on fertilizing capacity of sperm from fertile men.[55]

In summary one can see many discrepancies with respect to possible relaxin effects on male reproduction. Some disagreements can probably be explained by the variations in methodology and reflect variable functions of the relaxin in different species. These inconsistencies might in fact help to reveal the true biological role of relaxin signaling in normal development and disease. The potential involvement of relaxin in all these processes is intriguing at least.

Other Relaxin Peptides in Male Reproduction

Two other members of relaxin peptide family (*INLS5* and *INSL6*) are expressed in testis.[56,57] Both of them were initially identified through mining of the Genbank for the homology to relaxin peptide. As the other members of relaxin family, both of these genes are encoded by two small exons. *INSL6* is located in the close vicinity to relaxin genes in the genomes of mammalian species, indicating its possible origin through duplication events during evolution. In primates an additional member of this family, placenta-specific *INSL4*, is also present in the same genomic region along with two copies of relaxin gene (*RLN1* and *RLN2*).[58] The *INSL6* was found to be expressed at high levels in the testis as determined by Northern blot analysis and specifically within meiotic and post-meiotic germ cells in the seminiferous tubules as detected by in situ hybridization and immunohistochemistry analysis.[57] During mouse embryonic development the *Insl6* mRNA expression was detectable at low levels in testis until postnatal day 20 when there was a 40-50 fold increase in *Insl6* mRNA with appearance of pachytene spermatocytes, which were the major site of *Insl6* expression.[57] It was established that Insl6 is a secreted peptide, with structure containing disulfide bonds; it undergoes posttranslational modification which includes N-glycosylation, ubiquitination and cleavage by furin.[59] The receptor for INSL6 is not known, INSL6 does not activate G protein-coupled receptors of other relaxin family peptides.[60,61] Currently, there are no data indicating on any biological role or function of this peptide in testis.

The weak expression of INSL5 peptide was also detected in testis.[56] The secondary structure of the mature peptide was recently established. INSL5 was shown to be high affinity agonist for GPCR142 in vitro, it does not activate either LGR8 or LGR7, but has some weak antagonistic activity towards GPCR132 (relaxin 3/INSL7 receptor).[61] In functional guanosine (γ-thio)-triphosphate binding and cAMP accumulation assays, INSL5 potently activates GPCR142 and stimulates Ca^{2+} mobilization in HEK293 cells expressing GPCR142 and $G\alpha_{16}$.[62] Although expression of INSL5 and GPCR142 overlaps in several organs, their involvement in any physiological processes is currently unknown.

Insulin-Like Peptide 3, INSL3

Testicular Descent in Mammals

The INSL3 signaling plays a crucial role in the process of testicular descent in mammalian development.[63,64] The different position of the adult gonads is a distinctive feature of sexual dimorphism in the majority of mammals. Both testis and ovary are originated from the undifferentiated gonads. In females, the developing ovary remains in the high abdominal position. After testis differentiation, initiated by Y-chromosomal SRY gene, the developing testis migrates from its original abdominal position into the scrotum. The process of testis migration is called testicular descent (TD). It is believed that the low-temperature environment provided by such an extra-abdominal placement of testis is necessary for normal spermatogenesis. Despite notable differences in the anatomy of TD between mammals, the two stage model of TD in mammals was proposed separating two major phases—transabdominal and inguinoscrotal descent (Fig. 2).[65] It should be pointed out however, that up to five anatomically recognized stages of TD were defined based on the detailed histological analysis of the human embryos.[66] Two mesentery ligaments, attached to the developing gonad, the cranial suspensory (CSL) and gubernacular ligaments, are believed to play a major role in TD.[67] During the transabdominal phase between 10 and 23 weeks of gestation in human embryos (in mouse—between embryonic day 15.5 (15.5E) and 17.5E the testes gradually move from their original position in the urogenital ridge to the inguinal region and by birth to a bladder neck. High

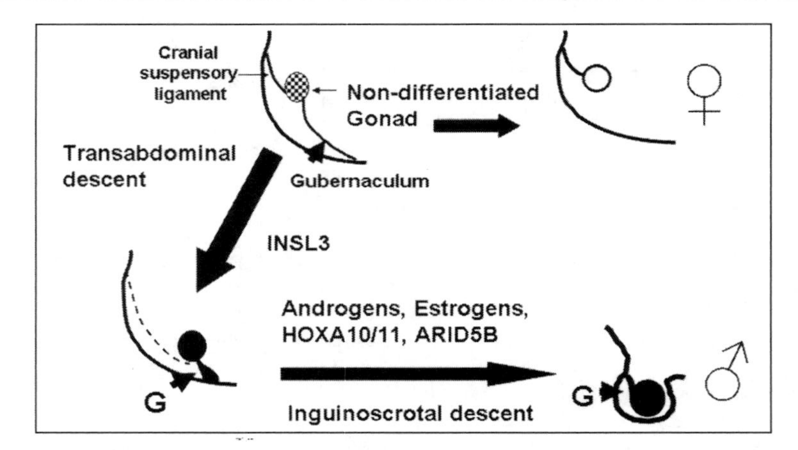

Figure 2. Two stages of testicular descent. The nondifferentiated gonad is located near kidney attached to the body wall with cranial suspensory ligament and inguinoscrotal ligament or gubernaculum (G). Persistence of cranial suspensory ligaments and the regression of gubernacula lead to the abdominal position of the ovary. The INSL3 peptide produced in testicular Leydig cells induces differentiation of gubernacula and causes first transabdominal stage of testicular descent into low abdominal position. During the second, inguinoscrotal phase, the testis migrate into the scrotum under control of hormonal (androgens, estrogens and perhaps INSL3) and genetic (homeodomain box genes (HOXA10, HOXA11), AT rich interactive domain 5B (ARID5B), etc) factors.

position of the ovary, attached to the abdominal wall through cranial ligament, seems to follow the default pathway. The process of transabdominal descent occurs in parallel with the shortening of gubernacular ligament cord, an outgrowth of the gubernacular bulb, differentiation and eversion of the cremaster muscle. In human fetus the second phase occurs before birth at weeks 24-34 of gestation, whereas in mice it is completed within 19 days after birth. This stage is characterized by the caudal extension of the gubernaculum, its involution and protrusion into the scrotal sac, development of the processus vaginalis, dilation of the inguinal canal by the gubernacular bulb and some intraabdominal pressure to force the testis through the canal. In mice, TD is mediated by the contractions of the cremaster muscle.[67] Human gubernacular histology evolves from a hydrated structure with a loose extracellular matrix and poorly differentiated fibroblasts into an essentially fibrous structure rich in collagen and elastic fibers. In rodents however, the adult gubernaculum is differentiated into cremaster muscle component of the spermatic cord.[68]

The cryptorchidism (from "Hidden Testis" in Greek) which in most cases is caused by extrascrotal testis position, is the most common congenital abnormality in newborn boys with a reported frequency of 2-4%.[69] There is a clear prevalence of the disease in premature born children, which in most cases is spontaneously corrected. By the age of one year the frequency of testicular maldescent is reduced to 1%.[69,70] Cryptorchidism is an established risk factor for infertility and testicular cancer and thus requires surgical intervention.

Genetic Ablation of Insl3 and Lgr8 Causes Cryptorchidism in Mice

Genetic targeting of Insl3 gene in mice resulted in bilateral cryptorchidism with testes located in high intraabdominal position.[71,72] Heterozygous animals had normally descended testes.[71,73] Testes and genital tracts in adult *Insl3*-deficient mice were mobile within the abdominal cavity, indicating the lack of the caudal attachment of the testis to inguinal region via gubernaculum with a regression of cranial attachments. Consequently, torsion of the vas deferens and testicular artery occurred in some mutant males. No signs of testicular malignancy were found in cryptorchid testis of *Insl3*-deficient mice (our unpublished results). Histological analysis revealed that the cryptorchidism in the mutant mice is due to the impairment of the gubernacular differentiation during the prenatal development. Proliferation and differentiation of the mesenchymal cells to myoblasts in outer layers of the gubernacular bulb is arrested in the *Insl3*-deficient embryos.[71,73] Thus, it was suggested that Insl3 induces the outgrowth of the gubernaculum during testicular descent.

The identification of the INSL3 receptor came once again from the studies of a mouse mutant. The *crsp* (*cryptorchidism with white spotting*) mutation was discovered among transgenic mice carrying a mouse tyrosinase minigene.[74] Homozygous *crsp/crsp* males exhibited the same phenotype as *Insl3*-deficient mice—high intraabdominal cryptorchidism associated with complete sterility. Remarkably, the mutant males were normally virulized, with a peripheral blood testosterone levels within the normal range. Similar to *Insl3*-deficiency, the *crsp* mutation did not affect spermatogenesis directly. Surgical removal of testis from the abdominal cavity resulted in a full restoration of spermatogenesis in the mutants.[71,74,75] Analysis of the genomic rearrangements in *crsp/crsp* mutant mice revealed a presence of a 550 kb deletion in the distal part of chromosome 5. Sequencing of this region lead to an identification of the novel gene encoding a protein with homology to the seven-transmembrane G protein-coupled receptors (GPCR). Presence of ten leucine-rich repeats (LRR) in the putative extracellular domain of the protein indicated that it belonged to the same family of GPCRs as glycoprotein hormone receptors. The highest degree of homology was detected to another orphan GPCR, LGR7. Genes encoding LGR7 and the novel receptor had the same exon-intron structure; corresponding peptides had the same domain organization; both contained the same number of LRRs and, a unique for this group, LDL class A receptor domain at their N-terminus. The expression of the novel gene named GREAT, *G protein-coupled receptor affecting testicular descent*, was detected in gubernaculum and testis implying its possible involvement in testicular descent. Subsequently, mice deficient for *Great* (later renamed *Lgr8* and *Rxfp2*) have been produced.[76] They manifested the same testis phenotype as the original *crsp* mice; thus confirming the role of *Lgr8* in testicular descent. Based on the phenotype similarity of the *Insl3*- and *Lgr8*-deficient mice, gene expression profile and the characteristics of the encoded polypeptide, it was suggested that LGR8 might be the cognate receptor for INSL3.[74]

Generation and characterization of transgenic mice with overexpression of Insl3 has further confirmed the role of this peptide in gubernacular differentiation. The *Insl3* transgene was expressed either specifically in the pancreas under rat insulin 2 promoter or ubiquitously in all tissues during pre- and postnatal development both in males and females.[77,78] Expression of insulin promoter-driven transgenic allele had rescued the testicular descent in Insl3-deficient male mice, indicating that the islet β-cells efficiently processed the *Insl3* product into the functional hormone.[78] Furthermore, the secretion of Insl3 from the pancreas and its effect on the developing gubernaculum clearly showed that Insl3 exerts its biological activity via an endocrine route.[78,79] All *Insl3*-transgenic females displayed bilateral inguinal hernia. The ovaries of transgenic animals descended into a position close to the bladder and were attached to the abdominal wall via the well-developed gubernaculum and cranial suspensory ligament. Importantly, the lack of Wolffian duct derivatives and development of the Müllerian duct derivatives indicated the absence of the androgen- and AMH-mediated activities in transgenic females. The gubernacular differentiation in the transgenic females precludes, therefore, a role of the androgens or AMH in TD and Insl3 is the only factor that is sufficient for an induction of gubernacular differentiation.

Injection of pregnant *Insl3*-transgenic animals with dihydrotestosterone results in the regression of CSL in transgenic embryos and thereby allowed further descent of the ovaries into the processus vaginalis.[78] These observations further support the hypothesis that the persistence of the fetal CSL may be an etiologic factor for some cases of cryptorchidism. Additionally, the processus vaginalis development in the *Insl3*-transgenic females rules out the involvement of the androgen-mediated activity in gubernacular evagination.[80,81] These findings are consistent with an inguinal hernia present in human male patients with complete androgen-insensitivity syndrome.[82,83]

The structure of the INSL3 peptide is homologous to relaxin. Relaxin was the first mammalian peptide hormone identified in almost 80 years ago. Despite intensive research the identity of the relaxin receptor remained elusive. Based on the mouse genetic studies described above Dr. A.

Hsueh and his colleagues suggested that relaxin signals through LGR7 and LGR8.[84] Using cells transfected with LGR7 or LGR8, they showed that treatment with native porcine relaxin caused a dose-dependent increase in intracellular cAMP concentration. These findings indicated that relaxin acted in vitro as a ligand for both GPCRs activating adenylate cyclase through G_s proteins.[84] It was shown later that human recombinant relaxin also stimulated both receptors; nevertheless, a significantly more efficient activation of LGR7 was noted.[85] It was demonstrated that the other members of insulin/relaxin family of peptides (INSL4, INSL5, INSL6, INSL7/RLN3) did not activate the LGR8 receptor.[60,84,86,87]

Synthetic INSL3 peptide selectively activates LGR8 but not LGR7 in vitro.[85] Similar to relaxin, the stimulation of the LGR8 receptor, transfected into 293 cells, with INSL3 causes a dose-dependent increase in the cAMP production. The direct interaction between INSL3 and LGR8 in a nanomolar concentration range was demonstrated by ligand receptor cross-linking indicating on a specific character of the ligand/receptor binding. It has been shown that the wild-type and AMH-deficient testes, which were cocultured with the gubernacular explants, had a growth stimulatory effect on gubernaculum. In contrast, testis from *Insl3*-deficient mice did not exhibit such activity. These results provide further evidence that INSL3 stimulates the outgrowth of the gubernaculum and exclude the role of AMH in this developmental process.[88] Similar results have been obtained in experiments with cultured gubernacular explants in medium supplemented with the recombinant INSL3 and synthetic androgen (R1881). Both hormones stimulated the outgrowth of the gubernacular explants to the same extent as wild-type control testis.[73] Treatment of cultured gubernacular cells with synthetic INSL3 stimulated cAMP production and cell proliferation.[85,89]

To analyze the Insl3/Lgr8 interaction in vivo we used transgenic mouse models.[60] As described above the transgenic overexpression of *Insl3* in females results in the outgrowth of gubernacula and descent of the ovaries into low abdominal position. We have shown that deficiency for the *Lgr8* gene in *Insl3* transgenic females completely abrogated such abnormal phenotype; they had a wild-type phenotype, with ovaries in the normal, high abdominal position; males of the same genotype developed cryptorchidism.[60] It was concluded therefore, that Lgr8 was the only cognate receptor for Insl3 in vivo. Deficiency for relaxin[33] or *Lgr7*,[34,35] as well as the transgenic overexpression of relaxin in mice[36] does not affect testicular descent. Thus, relaxin does not appear to affect LGR8 signaling in vivo and, therefore, it is not involved in testicular descent.

INSL3 and LGR8 Expression

Sexual dimorphic development of the gubernaculum coincides with the expression of INSL3 in prenatal Leydig cells of the testis. *Insl3* is first detected in murine testis at embryonic day 13.5 (E13.5) with a stable expression afterwards.[90,91] This coincides with the differentiation of fetal Leydig cells in mice. In contrast, no *Insl3* transcript could be detected in prenatal ovary. In male rat embryos, serum INSL3 concentrations are high at day 2 before parturition, decrease just before parturition and then, again, increase starting at day 10 after parturition until adult INSL3 concentrations are reached on day 39.[92] Such pattern of the *Insl3* expression corresponds to the timing of TD in rodents (Fig. 2).[93]

It was shown that Insl3 is expressed in the fetal population of Leydig cells in *hpg* mutant mouse.[94] These mice lack an active pituitary-gonadal axis, indicating that gonadotropins are not essential for the regulation of INSL3 expression during embryonic development. The prenatal exposure to 17 β-estradiol and the nonsteroidal synthetic estrogen diethylstilbestrol (DES) causes demasculinizing and feminizing effects in the male embryo, including impaired testicular descent (cryptorchidism). It was shown that maternal exposure to estrogens, including 17 α- and β-estradiol, as well as DES, down regulates *Insl3* expression in embryonic Leydig cells, thereby providing a mechanism for cryptorchidism.[95,96]

In adult testis the expression of INSL3 is constitutive and appears to reflect the differentiation status of adult Leydig cells. The treatment of mouse Leydig cells with different hormone or growth factors did not directly effect the *Insl3* expression.[97] In adult men, but not in women, the concentration of INSL3 in peripheral blood is high; it was demonstrated that the hormone is of testicular

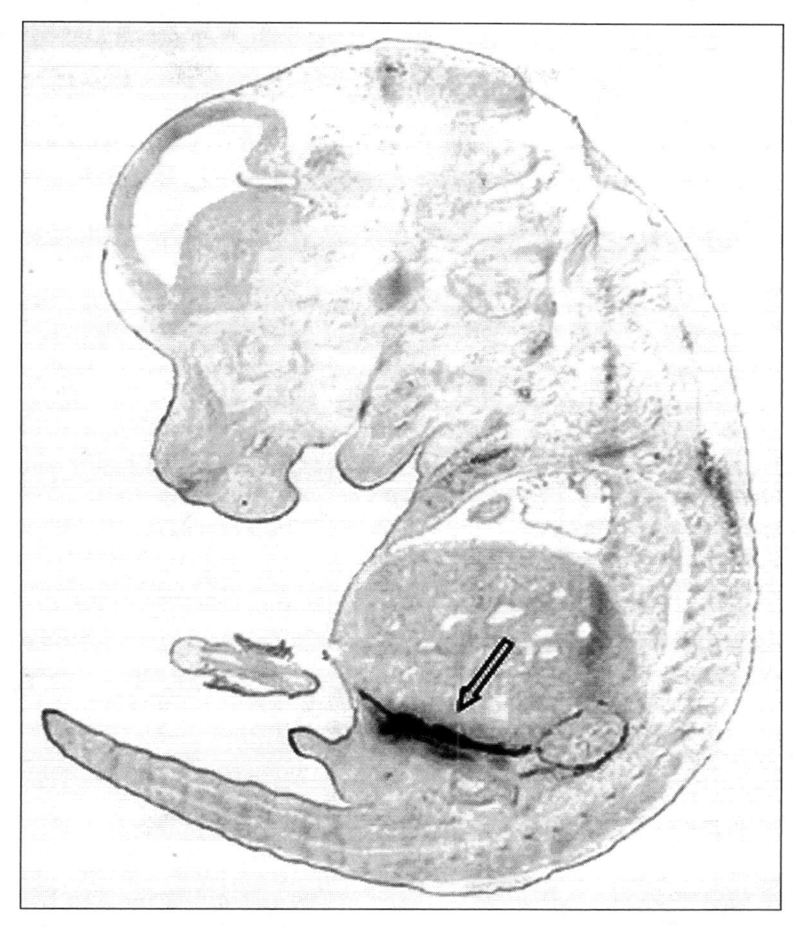

Figure 3. *Lgr8* expression in gubernaculum of 14.5 day male mouse embryo by in situ hybridization.[93]

origin.[98] It was hypothesized that decrease of INSL3 level might be associated with some forms of infertility and/or indicate the insufficient function of Leydig cells. Ferlin et al found a decrease in INSL3 level in severe oligospermic patients,[98] however this was not confirmed by Bay et al.[99] The differences may be explained by heterogeneity of infertile patients groups or variations in some unidentified compounding factors. In general, however, the INSL3 secretion is dependent on the pituitary-testicular differentiation effects. The patients with hypogonadotropic hypogonadism (HH) have low INSL3 concentration, their treatment with hCG caused an increase of INSL3. Conversely, the GnRH analog treatment of HH men or cyproterone acetate or testosterone treatment of fertile men caused reduction of INSL3 serum level. In row deer the seasonal variations in LH level and the concomitant waves of Leydig cell differentiation in testis correlate with increase of INSL3 serum level.[100]

What are the factors which regulate an increase of Insl3 expression during male development? Interestingly, the *INSL3* gene in mammals is located within the last exon of *JAK3* kinase. The distance between the 5'-located *JAK3* exon and the first exon of *INSL3* occupied by the putative promoter region of *INSL3* is relatively small. The computer analysis of this sequence revealed potential binding sites for several transcription factor. It was demonstrated that SF-1 (NR5A1) upregulated the activity of rodent[97] and canine *INSL3*[101] promoters in vitro. In mutant mice with Leydig cell-specific deficiency of Sf-1, the testes are cryptorchid suggesting abnormal Insl3 expression.[102] We have shown that the SF-1 binding site is conserved in sequenced mammalian promoters. The known antagonist of SF-1 factor, DAX-1 inhibited Insl3 promoter activity.[103] Significantly, both SF-1 and DAX-1 are the important players in male differentiation and thus provide a link between TD and sex determination process. Another transcriptional factor, NR4A1, regulates the INSL3 transcription in Leydig cells. Chromatin immunoprecipitation assays revealed that endogenous NR4A1 binds to the proximal Insl3 promoter in vivo. NR4A1 was implicated in cAMP-induced *Insl3* transcription in Leydig cells.[104]

The pattern of *Lgr8* expression provided a strong support to the involvement of Insl3 signaling in TD. At day E14.5 the expression of *Lgr8* is most prominent in the gubernacular ligaments (Fig. 3).[105] Remarkably, even after transabdominal TD there is a steady increase in the Lgr8 expression in gubernacular ligaments. Immunostaining for Lgr8 is first detected in the differentiating myoblasts in the basis of gubernacular bulb, in an adults in cremaster muscle cells (Agoulnik, Feng, unpublished data). The strong Lgr8 expression was detected in Leydig cells, in pre- and postmeiotic spermatogenic cells and in epididymis epithelial cells.[106]

Strong Insl3/Lgr8 expression after transabdominal TD suggests an involvement in the second phase of TD descent. Combining mouse mutants of *Gnrhr* receptor or *Hoxa10* homeobox gene with transgenic *Insl3* overexpression in pancreas we demonstrated that while transabdominal TD could occur in such mice, the Insl3 expression does not rescue the cryptorchid phenotype.[107]

What is the INSL3 role in adult reproductive organs? The orchydopexy (the surgical correction of cryptorchidism) in *Insl3-/-* or *Lgr8-/-* males restored spermatogenesis and some *Insl3*-deficient males were fertile. Thus, the presence of Insl3 is not critical for testis function. Can Insl3 signaling play a supporting role in male reproduction? INSL3 treatment in the gonadotropin deprivation model decrease apoptosis of the germ cells,[108] suggesting that this hormone may be cell survival factor. Interestingly, that in contrast to the stimulatory effects of INSL3 on cAMP production by gubernacular cells, LGR8 expressed in germ cells is coupled to the *Gi* protein, leading to a decrease of cAMP production. Thus, the INSL3 cellular action may be cell/tissue-specific.

Mutation Analysis of INSL3/LGR8 in Human Patients with Testicular Maldescent

The discovered effects of Insl3/Lgr8-deficiency on cryptorchidism in transgenic mutant mice stimulated a search for the mutant alleles of these genes in human patients with testicular maldescent. The analysis of sporadic and familial cases indeed revealed several mutant alleles in both genes. Most of the mutant alleles, caused by a single nucleotide/amino acid substitution, were found both in controls and in affected patients; some of them, however, were detected exclusively in the affected group. Surprisingly, all the patients with the INSL3/LGR8 mutations were heterozygous for the wild-type allele (Fig. 4), implying that the potential mutation-dependent mechanism of testicular maldescent may be caused by reduced signaling.[64]

As the other members of insulin-relaxin family of peptides, INSL3 is translated as a pre-pro-hormone and contains signal sequence, B-chain, C-peptide and A-chain. Upon hormone maturation, signal sequence and C-peptide are excised, while A- and B-chains are assembled with two inter-chain and one intra-chain disulfide bonds. All but one variants of the gene are missense mutations, the only exception is the R73X mutation, resulting in the termination of translation and a putative expression of only B peptide of INSL3. Recently Del Borgo et al showed that B peptide of INSL3 is an antagonist for LGR8 receptor.[109] The R73X mutation was found in a patient with unilateral cryptorchidism.[110] However, the mutation does not seem to be explicitly associated with the abnormal phenotype, as at least one male relative with the same mutation did not have cryptorchidism at the time of examination. The other cryptorchidism-specific mutations include R102H, R102C, R105H substitutions in C-peptide described in patients with bi- or unilateral cryptorchidism.[111] Several additional mutations such as W69R, R72K and P93L were found in patients with testicular maldescent and in infertile patients, suggesting some role of INSL3 in testis

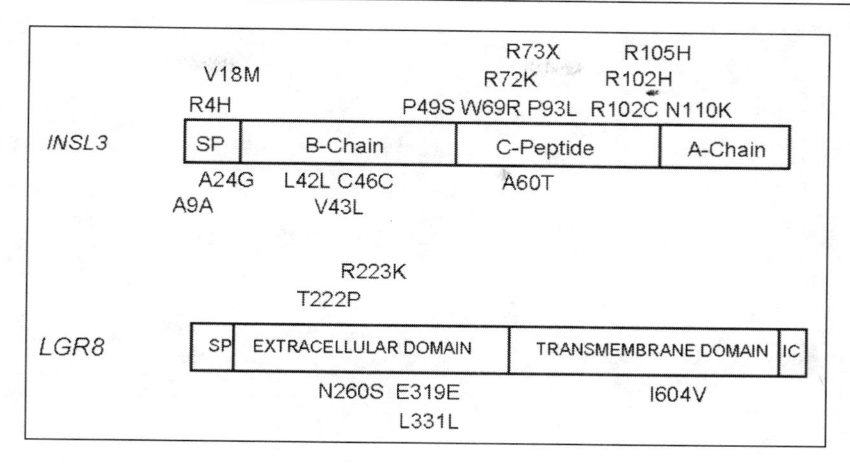

Figure 4. Mutations of the *INSL3* and *LGR8* genes found in human. The mutation shown above the protein were identified in patients with testicular maldescent; the mutations shown below the protein were found both in patient and control group. The blocks indicate the different protein domains of INSL3 and LGR8. SP, signal peptide; IC, intracellular domain.

function. A mutation in A-chain of INSL3 (N110K) was identified in a patient with the right testis located in the inguinal canal. N110 represents highly conservative residue, therefore a suggestion was made that N110K substitution would be deleterious for the function of INSL3.[112] Another study reported B-chain mutation (P49S) in a patient with bilateral intra-abdominal testes and undermasculinized genitalia. This patient had 46XY karyotype, female external genitalia but no uterus.[113] Functional analysis of the recombinant INSL3 peptides, containing all above-mentioned substitutions, revealed that only P49S mutation affects the ability of INSL3 to activate its cognate receptor.[60,111] The region of INSL3 B-chain surrounding P49 is involved in the interaction with the receptor,[114] which might explain the compromised properties of the mutant peptide. Further functional, population and, if possible, hereditary analysis of other INSL3 alleles is needed to ascertain the role of these mutation in the etiology of cryptorchidism.

The INSL3 receptor, LGR8, is a G protein-coupled receptor with a large extracellular N-terminus, intracellular C-terminus and a central part composed of 7 transmembrane regions. The extracellular domain of LGR8 includes ten leucine-rich repeats, believed to form highly organized structure participating in the ligand binding. T222P represents the only LGR8 mutation found so far, specific for the cryptorchid patients (Fig. 4).[76,115,116] The mutation was described in 24 patients with uni- and bilateral cryptorchidism; albeit some of the patients experienced spontaneous TD at puberty. Again, all patients contained this mutation in the heterozygous condition. The fact that individuals with similar mutations were presented with different clinical phenotype suggests that the other genetic and/or endocrine factors might affect the severity of INSL3-related deficiency. At the amino acid level the mutation affects one of the leucine-rich repeats and, according to functional analysis, renders the protein unable to express on the cell surface membrane (Fig. 5). Interestingly, all patients with T222P mutation were found in Italy and South of France. The haplotype analysis of the mutant allele indicates its common origin in this population.[115] Remarkably, in all cases where the parents of the affected patients were analyzed, the mutant allele was transmitted from the mother. The 100% association of the T222P mutation and testicular maldescent is highly significant for diagnostic purposes.

Summary and Future Directions

The available data indicate the importance of different peptides of the relaxin family in the development and anatomical position of the testis, the maintenance of spermatogenesis, as well as

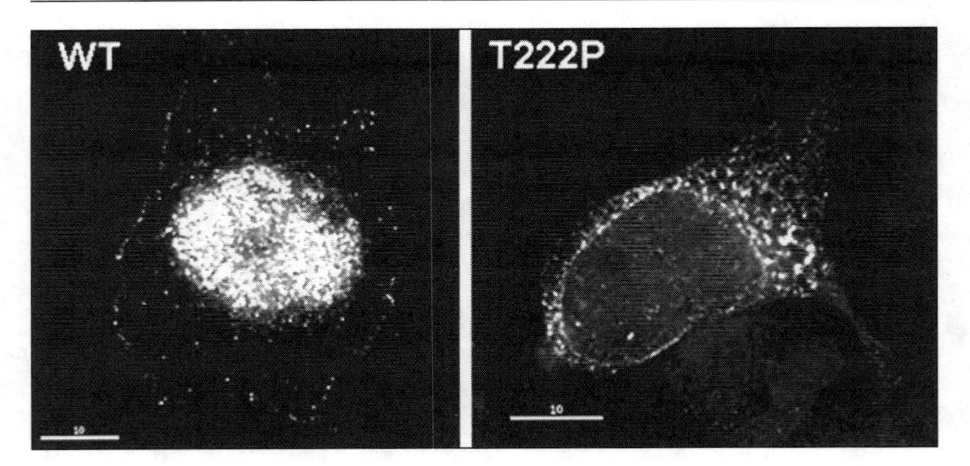

Figure 5. Mutant T222P LGR8 receptor fails to express on the cell surface membrane. Deconvolution microscopy of the 293T cells transfected with the expression constructs encoding wild-type (WT) and mutant T222P LGR8 receptor. The wild-type receptor is expressed on the cell surface seen as the outline of cell cytoplasm. Mutant T222P receptor is mainly located in the cytoplasm, seen as the uniform staining of the cell.

sperm functions. Relaxin appears to be mainly expressed in prostate in mammals and then secreted into seminal fluid; it stimulates and supports sperm motility and fertilization. It seems unlikely however that relaxin's role in male reproduction is essential as mutant mice with relaxin signaling deficiency are fertile. Further analysis of relaxin or relaxin antagonists are needed to assess their therapeutic significance in male infertility/contraceptive treatments. In the last 5-6 years significant attention has been devoted to another member of relaxin peptide family, the insulin-like 3 peptide. The product of Leydig cells, it controls the development of gubernacular ligaments and thus directs the descent of the testis during embryonic development into the scrotum. Mutations of INSL3 or its receptor cause cryptorchidism in mice and most likely in humans. The supporting role of this peptide in spermatogenesis is also reported. From the experimental perspective more efforts should be placed on identification of downstream targets of relaxin and INSL3 signaling. The strategic position of the INSL3 and relaxin G protein-coupled receptors on the cellular membrane and apparent endocrine function of the ligands makes these pairs an attractive target for drug design.

Acknowledgements

The author thanks the current and former members of the lab for their earnest contribution to the relaxin/INSL3 studies. I also thank Ms. Eva Agoulnik for editorial comments. Research in this laboratory was supported by grants from the National Institute of Child Health and Human Development and National Cancer Institute.

References

1. Weiss G. Relaxin in the male. Biol Reprod 1989; 40(2):197-200.
2. Kohsaka T, Takahara H, Sasada H et al. Evidence for immunoreactive relaxin in boar seminal vesicles using combined light and electron microscope immunocytochemistry. J Reprod Fertil 1992; 95(2):397-408.
3. Essig M, Schoenfeld C, D'Eletto RT et al. Relaxin in human seminal plasma. Ann N Y Acad Sci 1982; 380:224-230.
4. Weiss G, Goldsmith LT, Schoenfeld C et al. Partial purification of relaxin from human seminal plasma. Am J Obstet Gynecol 1986; 154(4):749-755.
5. Juang HH, Musah AI, Schwabe C et al. Relaxin in peripheral plasma of boars during development, copulation, after administration of hCG and after castration. J Reprod Fertil 1996; 107(1):1-6.

6. Sokol RZ, Wang XS, Lechago J et al. Immunohistochemical localization of relaxin in human prostate. J Histochem Cytochem 1989; 37(8):1253-1255.
7. Yki-Jarvinen H, Wahlstrom T, Seppala M. Immunohistochemical demonstration of relaxin in the genital tract of men. J Reprod Fertil 1983; 69(2):693-695.
8. Hansell DJ, Bryant-Greenwood GD, Greenwood FC. Expression of the human relaxin H1 gene in the decidua, trophoblast and prostate. Journal of Clinical Endocrinology and Metabolism 1991; 72(4):899-904.
9. Bogic LV, Mandel M, Bryant-Greenwood GD. Relaxin gene expression in human reproductive tissues by in situ hybridization. Journal of Clinical Endocrinology and Metabolism 1995; 80(1):130-137.
10. Gunnersen JM, Fu P, Roche PJ et al. Expression of human relaxin genes: characterization of a novel alternatively-spliced human relaxin mRNA species. Mol Cell Endocrinol 1996; 118(1-2):85-94.
11. Gunnersen JM, Roche PJ, Tregear GW et al. Characterization of human relaxin gene regulation in the relaxin-expressing human prostate adenocarcinoma cell line LNCaP.FGC. Journal of Molecular Endocrinology 1995; 15(2):153-166.
12. Garibay-Tupas JL, Bao S, Kim MT et al. Isolation and analysis of the 3'-untranslated regions of the human relaxin H1 and H2 genes. Journal of Molecular Endocrinology 2000; 24(2):241-252.
13. Winslow JW, Shih A, Bourell JH et al. Human seminal relaxin is a product of the same gene as human luteal relaxin. Endocrinology 1992; 130(5):2660-2668.
14. Hudson P, John M, Crawford R et al. Relaxin gene expression in human ovaries and the predicted structure of a human preprorelaxin by analysis of cDNA clones. EMBO Journal 1984; 3(10):2333-2339.
15. Tashima LS, Yamamoto SY, Yasuda M et al. Decidual relaxins: gene and protein up-regulation in preterm premature rupture of the membranes by complementary DNA arrays and quantitative immunocytochemistry. Am J Obstet Gynecol 2002; 187(3):785-797.
16. Anderson MB, Collado-Torres M, Vaupel MR. Absence of relaxin immunostaining in the male reproductive tracts of the rat and mouse. J Histochem Cytochem 1986; 34(7):945-948.
17. Gunnersen JM, Crawford RJ, Tregear GW. Expression of the relaxin gene in rat tissues. Mol Cell Endocrinol 1995; 110(1-2):55-64.
18. Samuel CS, Tian H, Zhao L et al. Relaxin is a key mediator of prostate growth and male reproductive tract development. Lab Invest 2003; 83(7):1055-1067.
19. Thompson VC, Morris TG, Cochrane DR et al. Relaxin becomes upregulated during prostate cancer progression to androgen independence and is negatively regulated by androgens. Prostate 2006; 66(16):1698-1709.
20. Garibay-Tupas JL, Okazaki KJ, Tashima LS et al. Regulation of the human relaxin genes H1 and H2 by steroid hormones. Mol Cell Endocrinol 2004; 219(1-2):115-125.
21. Dubois MP, Dacheux JL. Relaxin, a male hormone? Immunocytological localization of a related antigen in the boar testis. Cell Tissue Res 1978; 187(2):201-214.
22. Arakaki RF, Kleinfeld RG, Bryant-Greenwood GD. Immunofluorescence studies using antisera to crude and to purified porcine relaxin. Biol Reprod 1980; 23(1):153-159.
23. Bryant-Greenwood GD, Niall HD, Greenwood FC. Relaxin: proceedings of a Workshop on the Chemistry and Biology of Relaxin held at the East-West Center, the University of Hawaii, Honolulu, Hawaii, 1980. New York: Elsevier/North-Holland, 1981.
24. Parry LJ, Rust W, Ivell R. Marsupial relaxin: complementary deoxyribonucleic acid sequence and gene expression in the female and male tammar wallaby, Macropus eugenii. Biol Reprod 1997; 57(1):119-127.
25. de Rienzo G, Aniello F, Branno M et al. Isolation and characterization of a novel member of the relaxin/insulin family from the testis of the frog Rana esculenta. Endocrinology 2001; 142(7):3231-3238.
26. Hsu SY, Kudo M, Chen T et al. The three subfamilies of leucine-rich repeat-containing G protein-coupled receptors (LGR): identification of LGR6 and LGR7 and the signaling mechanism for LGR7. Mol Endocrinol 2000; 14(8):1257-1271.
27. Muda M, He C, Martini PG et al. Splice variants of the relaxin and INSL3 receptors reveal unanticipated molecular complexity. Mol Hum Reprod 2005; 11(8):591-600.
28. Scott DJ, Layfield S, Yan Y et al. Characterization of novel splice variants of LGR7 and LGR8 reveals that receptor signaling is mediated by their unique LDLa modules. J Biol Chem 2006.
29. Scott DJ, Tregear GW, Bathgate RA. LGR7-truncate is a splice variant of the relaxin receptor LGR7 and is a relaxin antagonist in vitro. Ann N Y Acad Sci 2005; 1041:22-26.
30. Min G, Sherwood OD. Localization of specific relaxin-binding cells in the ovary and testis of pigs. Biol Reprod 1998; 59(2):401-408.
31. Silvertown JD, Ng J, Sato T et al. H2 relaxin overexpression increases in vivo prostate xenograft tumor growth and angiogenesis. Int J Cancer 2006; 118(1):62-73.

32. Vinall RL, Tepper CG, Shi XB et al. The R273H p53 mutation can facilitate the androgen-independent growth of LNCaP by a mechanism that involves H2 relaxin and its cognate receptor LGR7. Oncogene 2006; 25(14):2082-2093.
33. Zhao L, Roche PJ, Gunnersen JM et al. Mice without a functional relaxin gene are unable to deliver milk to their pups. Endocrinology 1999; 140(1):445-453.
34. Kamat AA, Feng S, Bogatcheva NV et al. Genetic targeting of relaxin and insulin-like factor 3 receptors in mice. Endocrinology 2004; 145(10):4712-4720.
35. Krajnc-Franken MA, van Disseldorp AJ, Koenders JE et al. Impaired nipple development and parturition in LGR7 knockout mice. Mol Cell Biol 2004; 24(2):687-696.
36. Feng S, Bogatcheva NV, Kamat AA et al. Endocrine effects of relaxin overexpression in mice. Endocrinology 2006; 147(1):407-414.
37. Carrell DT, Peterson CM, Urry RL. The binding of recombinant human relaxin to human spermatozoa. Endocr Res 1995; 21(3):697-707.
38. Brenner SH, Lesing JB, Schoenfeld C et al. Human semen relaxin and its correlation with the parameters of semen analysis. Fertil Steril 1987; 47(4):714-716.
39. Schieferstein G, Voelter W, Seeger H et al. Immunoreactive relaxin in seminal plasma of man. Int J Fertil 1989; 34(3):215-218.
40. Sasaki Y, Kohsaka T, Kawarasaki T et al. Immunoreactive relaxin in seminal plasma of fertile boars and its correlation with sperm motility characteristics determined by computer-assisted digital image analysis. Int J Androl 2001; 24(1):24-30.
41. Kohsaka T, Hamano K, Sasada H et al. Seminal immunoreactive relaxin in domestic animals and its relationship to sperm motility as a possible index for predicting the fertilizing ability of sires. Int J Androl 2003; 26(2):115-120.
42. Neuwinger J, Jockenhovel F, Nieschlag E. The influence of relaxin on motility of human sperm in vitro. Andrologia 1990; 22(4):335-339.
43. Sarosi P, Schoenfeld C, Berman J et al. Effect of anti-relaxin antiserum on sperm motility in vitro. Endocrinology 1983; 112(5):1860-1861.
44. Fuchs U. The effect of relaxin and ubiquitin on sperm motility. Zentralbl Gynakol 1993; 115(3):117-120.
45. Jockenhovel F, Altensell A, Nieschlag E. Active immunization with relaxin does not influence objectively determined sperm motility characteristics in rabbits. Andrologia 1990; 22(2):171-178.
46. Essig M, Schoenfeld C, Amelar RD et al. Stimulation of human sperm motility by relaxin. Fertil Steril 1982; 38(3):339-343.
47. Lessing JB, Brenner SH, Schoenfeld C et al. The effect of relaxin on the motility of sperm in freshly thawed human semen. Fertil Steril 1985; 44(3):406-409.
48. Lessing JB, Brenner SH, Colon JM et al. Effect of relaxin on human spermatozoa. J Reprod Med 1986; 31(5):304-309.
49. Colon JM, Ginsburg F, Lessing JB et al. The effect of relaxin and prostaglandin E2 on the motility of human spermatozoa. Fertil Steril 1986; 46(6):1133-1139.
50. Han YJ, Miah AG, Yoshida M et al. Effect of relaxin on in vitro fertilization of porcine oocytes. J Reprod Dev 2006; 52(5):657-662.
51. Park JM, Ewing K, Miller F et al. Effects of relaxin on the fertilization capacity of human spermatozoa. Am J Obstet Gynecol 1988; 158(4):974-979.
52. Brenner SH, Lessing JB, Schoenfeld C et al. Stimulation of human sperm cervical mucus penetration in vitro by relaxin. Fertil Steril 1984; 42(1):92-96.
53. Pupula M, Quinn P, MacLennan A. The effect of porcine relaxin on the fertilisation of mouse oocytes in vitro. Clin Reprod Fertil 1986; 4(6):383-387.
54. Miah AG, Tareq KM, Hamano KI et al. Effect of Relaxin on Acrosome Reaction and Utilization of Glucose in Boar Spermatozoa. J Reprod Dev 2006; 52(6):773-779.
55. Chan SYW, Tang LCH. Lack of effect of exogenous relaxin on the fertilizing capacity of human spermatozoa. IRCS Med Sci 1984; 12:879-880.
56. Conklin D, Lofton-Day CE, Haldeman BA et al. Identification of INSL5, a new member of the insulin superfamily. Genomics 1999; 60(1):50-56.
57. Lok S, Johnston DS, Conklin D et al. Identification of INSL6, a new member of the insulin family that is expressed in the testis of the human and rat. Biol Reprod 2000; 62(6):1593-1599.
58. Chassin D, Laurent A, Janneau JL et al. Cloning of a new member of the insulin gene superfamily (INSL4) expressed in human placenta. Genomics 1995; 29(2):465-470.
59. Lu C, Walker WH, Sun J et al. Insulin-Like Peptide 6: Characterization of Secretory Status and Post-translational Modifications. Endocrinology 2006; 147(12):5611-5623.
60. Bogatcheva NV, Truong A, Feng S et al. GREAT/LGR8 is the only receptor for insulin-like 3 peptide. Mol Endocrinol 2003; 17(12):2639-2646.

61. Liu C, Chen J, Kuei C et al. Relaxin-3/insulin-like peptide 5 chimeric peptide, a selective ligand for G protein-coupled receptor (GPCR)135 and GPCR142 over leucine-rich repeat-containing G protein-coupled receptor 7. Mol Pharmacol 2005; 67(1):231-240.
62. Liu C, Kuei C, Sutton S et al. INSL5 is a high affinity specific agonist for GPCR142 (GPR100). J Biol Chem 2005; 280(1):292-300.
63. Adham IM, Agoulnik AI. Insulin-like 3 signalling in testicular descent. Int J Androl 2004; 27(5):257-265.
64. Bogatcheva NV, Agoulnik AI. INSL3/LGR8 role in testicular descent and cryptorchidism. Reprod Biomed Online 2005; 10(1):49-54.
65. Hutson JM. A biphasic model for the hormonal control of testicular descent. Lancet 1985; 2(8452):419-421.
66. Barteczko KJ, Jacob MI. The testicular descent in human. Origin, development and fate of the gubernaculum Hunteri, processus vaginalis peritonei and gonadal ligaments. Adv Anat Embryol Cell Biol 2000; 156:III-X, 1-98.
67. Hutson JM, Hasthorpe S, Heyns CF. Anatomical and functional aspects of testicular descent and cryptorchidism. Endocr Rev 1997; 18(2):259-280.
68. Hrabovszky Z, Di Pilla N, Yap T et al. Role of the gubernacular bulb in cremaster muscle development of the rat. Anat Rec 2002; 267(2):159-165.
69. Thonneau PF, Gandia P, Mieusset R. Cryptorchidism: incidence, risk factors and potential role of environment; an update. J Androl 2003; 24(2):155-162.
70. Morley R, Lucas A. Undescended testes in low birthweight infants. Br Med J (Clin Res Ed) 1987; 295(6601):753.
71. Zimmermann S, Steding G, Emmen JM et al. Targeted disruption of the Insl3 gene causes bilateral cryptorchidism. Mol Endocrinol 1999; 13(5):681-691.
72. Nef S, Parada LF. Cryptorchidism in mice mutant for Insl.3 Nat Genet 1999; 22(3):295-299.
73. Kubota Y, Temelcos C, Bathgate RA et al. The role of insulin 3, testosterone, Mullerian inhibiting substance and relaxin in rat gubernacular growth. Mol Hum Reprod 2002; 8(10):900-905.
74. Overbeek PA, Gorlov IP, Sutherland RW et al. A transgenic insertion causing cryptorchidism in mice. Genesis 2001; 30(1):26-35.
75. Nguyen MT, Showalter PR, Timmons CF et al. Effects of orchiopexy on congenitally cryptorchid insulin-3 knockout mice. J Urol 2002; 168(4 Pt 2):1779-1783.
76. Gorlov IP, Kamat A, Bogatcheva NV et al. Mutations of the GREAT gene cause cryptorchidism. Hum Mol Genet 2002; 11(19):2309-2318.
77. Koskimies P, Suvanto M, Nokkala E et al. Female mice carrying a ubiquitin promoter-Insl3 transgene have descended ovaries and inguinal hernias but normal fertility. Mol Cell Endocrinol 2003; 206(1-2):159-166.
78. Adham IM, Steding G, Thamm T et al. The overexpression of the insl3 in female mice causes descent of the ovaries. Mol Endocrinol 2002; 16(2):244-252.
79. Bullesbach EE, Rhodes R, Rembiesa B et al. The relaxin-like factor is a hormone. Endocrine 1999; 10(2):167-169.
80. Wensing CJ. The embryology of testicular descent. Horm Res 1988; 30(4-5):144-152.
81. Wensing CJ, Colenbrander B. Normal and abnormal testicular descent. Oxf Rev Reprod Biol 1986; 8:130-164.
82. Heyns CF, Hutson JM. Historical review of theories on testicular descent. J Urol 1995; 153(3 Pt 1):754-767.
83. Hutson JM, Beasley SW. Embryological controversies in testicular descent. Semin Urol 1988; 6(2):68-73.
84. Hsu SY, Nakabayashi K, Nishi S et al. Activation of orphan receptors by the hormone relaxin. Science 2002; 295(5555):671-674.
85. Kumagai J, Hsu SY, Matsumi H et al. INSL3/Leydig insulin-like peptide activates the LGR8 receptor important in testis descent. J Biol Chem 2002; 277(35):31283-31286.
86. Sudo S, Kumagai J, Nishi S et al. H3 relaxin is a specific ligand for LGR7 and activates the receptor by interacting with both the ectodomain and the exoloop 2. J Biol Chem 2003; 278(10):7855-7862.
87. Feng S, Bogatcheva NV, Kamat AA et al. Genetic targeting of relaxin and insl3 signaling in mice. Ann N Y Acad Sci 2005; 1041:82-90.
88. Emmen JM, McLuskey A, Adham IM et al. Hormonal control of gubernaculum development during testis descent: gubernaculum outgrowth in vitro requires both insulin-like factor and androgen. Endocrinology 2000; 141(12):4720-4727.
89. Hsu SY, Nakabayashi K, Nishi S et al. Relaxin signaling in reproductive tissues. Mol Cell Endocrinol 2003; 202(1-2):165-170.

90. Pusch W, Balvers M, Ivell R. Molecular cloning and expression of the relaxin-like factor from the mouse testis. Endocrinology 1996; 137(7):3009-3013.
91. Zimmermann S, Schottler P, Engel W et al. Mouse Leydig insulin-like (Ley I-L) gene: structure and expression during testis and ovary development. Mol Reprod Dev 1997; 47(1):30-38.
92. Boockfor FR, Fullbright G, Bullesbach EE et al. Relaxin-like factor (RLF) serum concentrations and gubernaculum RLF receptor display in relation to pre- and neonatal development of rats. Reproduction 2001; 122(6):899-906.
93. Visel A, Thaller C, Eichele G. GenePaint.org: an atlas of gene expression patterns in the mouse embryo. Nucleic Acids Res 2004; 32(Database issue):D552-556.
94. Balvers M, Spiess AN, Domagalski R et al. Relaxin-like factor expression as a marker of differentiation in the mouse testis and ovary. Endocrinology 1998; 139(6):2960-2970.
95. Emmen JM, McLuskey A, Adham IM et al. Involvement of insulin-like factor 3 (Insl3) in diethylstilbestrol-induced cryptorchidism. Endocrinology 2000; 141(2):846-849.
96. Nef S, Shipman T, Parada LF. A molecular basis for estrogen-induced cryptorchidism. Dev Biol 2000; 224(2):354-361.
97. Sadeghian H, Anand-Ivell R, Balvers M et al. Constitutive regulation of the Insl3 gene in rat Leydig cells. Mol Cell Endocrinol 2005; 241(1-2):10-20.
98. Foresta C, Bettella A, Vinanzi C et al. A novel circulating hormone of testis origin in humans. J Clin Endocrinol Metab 2004; 89(12):5952-5958.
99. Bay K, Hartung S, Ivell R et al. Insulin-like factor 3 serum levels in 135 normal men and 85 men with testicular disorders: relationship to the luteinizing hormone-testosterone axis. J Clin Endocrinol Metab 2005; 90(6):3410-3418.
100. Hombach-Klonisch S, Schon J, Kehlen A et al. Seasonal expression of INSL3 and Lgr8/Insl3 receptor transcripts indicates variable differentiation of Leydig cells in the roe deer testis. Biol Reprod 2004; 71(4):1079-1087.
101. Truong A, Bogatcheva NV, Schelling C et al. Isolation and expression analysis of the canine insulin-like factor 3 gene. Biol Reprod 2003; 69(5):1658-1664.
102. Jeyasuria P, Ikeda Y, Jamin SP et al. Cell-specific knockout of steroidogenic factor 1 reveals its essential roles in gonadal function. Mol Endocrinol 2004; 18(7):1610-1619.
103. Koskimies P, Levallet J, Sipila P et al. Murine relaxin-like factor promoter: functional characterization and regulation by transcription factors steroidogenic factor 1 and DAX-1. Endocrinology 2002; 143(3):909-919.
104. Robert NM, Martin LJ, Tremblay JJ. The orphan nuclear receptor NR4A1 regulates insulin-like 3 gene transcription in Leydig cells. Biol Reprod 2006; 74(2):322-330.
105. Agoulnik AI. Mouse mutants of relaxin, insulin-like 3 peptide and their receptors. Curr Med Chem Immunol Endocr Metab Agents 2005; 5(5):411-419.
106. Anand-Ivell RJ, Relan V, Balvers M et al. Expression of the insulin-like peptide 3 (INSL3) hormone-receptor (LGR8) system in the testis. Biol Reprod 2006; 74(5):945-953.
107. Feng S, Bogatcheva NV, Truong A et al. Over expression of insulin-like 3 does not prevent cryptorchidism in GNRHR or HOXA10 deficient mice. J Urol 2006; 176(1):399-404.
108. Kawamura K, Kumagai J, Sudo S et al. Paracrine regulation of mammalian oocyte maturation and male germ cell survival. Proc Natl Acad Sci USA 2004; 101(19):7323-7328.
109. Del Borgo MP, Hughes RA, Bathgate RA et al. Analogs of insulin-like peptide 3 (INSL3) B-chain are LGR8 antagonists in vitro and in vivo. J Biol Chem 2006; 281(19):13068-13074.
110. Tomboc M, Lee PA, Mitwally MF et al. Insulin-like 3/relaxin-like factor gene mutations are associated with cryptorchidism. J Clin Endocrinol Metab 2000; 85(11):4013-4018.
111. Ferlin A, Bogatcheva NV, Gianesello L et al. Insulin-like factor 3 gene mutations in testicular dysgenesis syndrome: clinical and functional characterization. Mol Hum Reprod 2006; 12(6):401-406.
112. Canto P, Escudero I, Soderlund D et al. A novel mutation of the insulin-like 3 gene in patients with cryptorchidism. J Hum Genet 2003; 48(2):86-90.
113. Lim HN, Raipert-de Meyts E, Skakkebaek NE et al. Genetic analysis of the INSL3 gene in patients with maldescent of the testis. Eur J Endocrinol 2001; 144(2):129-137.
114. Bullesbach EE, Schwabe C. Tryptophan B27 in the relaxin-like factor (RLF) is crucial for RLF receptor-binding. Biochemistry 1999; 38(10):3073-3078.
115. Ferlin A, Simonato M, Bartoloni L et al. The INSL3-LGR8/GREAT ligand-receptor pair in human cryptorchidism. J Clin Endocrinol Metab 2003; 88(9):4273-4279.
116. Bogatcheva NV, Ferlin A, Feng S et al. T222P mutation of the insulin-like 3 hormone receptor LGR8 is associated with testicular maldescent and hinders receptor expression on the cell surface membrane. Am J Physiol Endocrinol Metab 2007; 292(1):138-144.

Chapter 6

The Vascular Actions of Relaxin

Arundhathi Jeyabalan, Sanjeev G. Shroff, Jaqueline Novak and Kirk P. Conrad*

Abstract

Relaxin is emerging as a hormone with important vascular actions. Much of our recently gained knowledge of relaxin in this context has stemmed from investigations of maternal vascular adaptations to pregnancy in which the hormone is turning out to be an important mediator. This chapter is separated into three parts. In Part 1, we discuss relaxin in the setting of normal vascular function and focus on systemic hemodynamics and arterial mechanical properties, renal and other peripheral circulations, angiogenesis, as well as the cellular mechanisms of the vasodilatory actions of relaxin. In this section, we also summarize the evidence for an arterial-derived relaxin ligand-receptor system. In Part 2, we present relaxin in the context of vascular dysfunction and the implications for relaxin as a therapeutic agent in renal and cardiac diseases, ischemia and reperfusion injury, pulmonary hypertension, vascular inflammation and preeclampsia. Finally, in Part 3, we highlight some of the controversies and unresolved issues, as well as suggest a general direction for future relaxin research that is urgently needed.

Introduction

The discovery that the vasculature may be another target of relaxin was made by Frederick L. Hisaw and colleagues. They reported that following administration of relaxin (Rlx) to castrated monkeys, there were profound morphological alterations in the endothelial cells of the endometrium consistent with hypertrophy and hyperplasia, as well as enlargement of arterioles and capillaries.[1,2] The concept that relaxin alters vascular structure and function has been subsequently bolstered by numerous investigations, particularly in the last decade. Much of our recent understanding of relaxin as a vascular hormone has arisen from studies of maternal renal and cardiovascular adaptations to pregnancy in which relaxin is emerging as a pivotal player. The objective of this chapter is to review the vascular actions of relaxin.

Part 1. Contribution of Relaxin to Normal Vascular Function (Table 1)

Influence of Relaxin on Systemic Hemodynamics and Arterial Mechanical Properties

Definitions

Systemic arterial load is defined as the mechanical opposition to the flow of blood out of the left ventricle.[3] There are 2 components. The first is the steady arterial load commonly known as systemic vascular resistance (SVR), which is calculated by the quotient of mean arterial pressure and cardiac output (CO) and results mainly from arteriolar properties. The second is pulsatile arterial

*Corresponding Author: Kirk P. Conrad—Department of Physiology and Functional Genomics, University of Florida College of Medicine, Gainesville, Florida 32610-0274, U.S.A.
Email: kpconrad@ufl.edu

Relaxin and Related Peptides, edited by Alexander I. Agoulnik. ©2007 Landes Bioscience and Springer Science+Business Media.

Table 1. Summary of the contribution of relaxin to normal vascular function

I.	**Systemic Hemodynamics and Arterial Mechanical Properties** • RLX administration increases cardiac output and arterial compliance, and reduces systemic vascular resistance and myogenic reactivity in rats. • Endogenous, circulating relaxin mediates the increased cardiac output and arterial compliance, and the reduced systemic vascular resistance and myogenic reactivity during mid-term pregnancy in rats.
II.	**Renal Circulation** • RLX administration reduces renal vascular resistance and increases renal plasma flow and glomerular filtration rate in rats and humans. • Endogenous, circulating relaxin mediates the reduced renal vascular resistance and increased renal plasma flow and glomerular filtration rate during pregnancy in rats. • Endogenous, circulating relaxin contributes to increased glomerular filtration rate during pregnancy in women.
III.	**Other Organ Circulations** • RLX administration increases blood flow in other organ circulations: coronary, uterus, mammary gland, liver, mesentery and mesocaecum.
IV.	**Angiogenesis** • RLX has been shown to be angiogenic in the endometrium, and in the setting of wound healing and myocardial infarction.
V.	**Local Vascular Relaxin Ligand-Receptor System** • Recent evidence indicates the local expression of relaxin ligand and receptor in arteries that increases arterial compliance and reduces myogenic reactivity.

load, which becomes relevant because of the inherently pulsatile nature of the cardiac pump and is determined by vessel geometry and wall visco-elasticity, the branching of the vasculature that yields wave propagations and reflections and the mechanical attributes of blood. Together, the steady and pulsatile arterial loads constitute a comprehensive characterization of total hydraulic load in terms of arterial mechanical properties. Global arterial compliance (global AC) is one measure of pulsatile arterial load, which is derived from CO and the diastolic decay of the aortic pressure waveform or more simply from the ratio of the stroke volume and pulse pressure.

Systemic Hemodynamics and Arterial Mechanical Properties during Pregnancy

A fundamental cardiovascular adaptation to human pregnancy is the increase in global AC which reaches a peak by the end of the first or beginning of the second trimester just as SVR reaches a nadir.[4] At least in theory, the increase in global AC is essential to the maintenance of cardiovascular homeostasis during pregnancy for several reasons: (i) by preventing an excessive fall in diastolic pressure which otherwise would decline to precariously low levels due to the large drop in SVR; (ii) by minimizing the pulsatile or oscillatory work (i.e., wasted energy) which otherwise would rise disproportionately to the increase in total work required of and expended by the heart during pregnancy; (iii) by contributing to arterial underfilling along with the reduction in SVR (albeit to a lesser degree than SVR), both of which abet renal sodium and water retention and plasma volume expansion during early pregnancy; and (iv) by preserving steady shear relative to steady shear stress at the blood-endothelial interface in the setting of the hyperdynamic circulation of pregnancy, thus favoring nitric oxide production by the endothelium over that of superoxide and other potentially damaging reactive oxygen species.

Influence of Relaxin Administration on Systemic Hemodynamics and Arterial Mechanical Properties in Nonpregnant Females and Males

Because relaxin mediates the maternal renal circulatory changes during pregnancy in rats,[5] it was logical to consider whether the hormone might also underlie the broader cardiovascular adaptations to pregnancy, i.e., the elevations in CO and global AC, as well as the decline in SVR. To begin addressing this question, Conrad and coworkers first took a pharmacological approach and investigated whether recombinant human relaxin (rhRLX) has the potential to modify systemic cardiovascular function by chronically administering the hormone to nonpregnant female rats.[6] A methodology was developed to measure global AC in conscious, unrestrained rats. rhRLX was infused subcutaneously for 10 days and circulating levels were similar to those seen in rats during early to midterm pregnancy. Significant increases in CO, global AC, as well as reductions in SVR were noted by the earliest time point of measurement (day 2 or 3) and they were maintained throughout the 10 days of rhRLX infusion, the overall magnitude of these changes being ~ 20%. However, mean arterial pressure was unchanged throughout the 10-day hormone infusion, because the fall in SVR was matched by a compensatory increase in CO, the latter mainly due to augmentation of stroke volume. Finally, small renal arteries dissected from female rats after 5 days of rhRLX administration and subsequently mounted in a pressure arteriograph and treated with papaverine and EGTA to block smooth muscle function, exhibited significant increases in compliance in comparison to those arteries isolated from vehicle treated animals.[6] Recent evidence indicates that, in addition to large arteries, small arteries also make an important contribution to the global AC.[7] These results suggest that, in addition to the decrease in vascular smooth muscle tone due to the vasodilatory attribute of relaxin, the increase in global AC observed in vivo was also a consequence of alterations in vascular structure, i.e., cellular components or extracellular matrix in the blood vessel wall. However, the precise nature of the changes in vascular structure contributing to the increase in arterial compliance during rhRLX administration to nonpregnant rats remains to be determined.

Chronic administration of rhRLX over days to either normotensive control or hypertensive, male and female rats mimicked the alterations in systemic hemodynamic and arterial mechanical properties of gestation.[6,8,9] In contrast, the short-term administration of rhRLX over hours was only effective in the angiotensin-II model of hypertension, but not in spontaneously hypertensive or normotensive rats.[8,9] Thus, relaxin apparently acts more rapidly in the renal circulation of normotensive rats than in the systemic vasculature (see below Influence of relaxin administration on the renal circulation in nonpregnant females and males).

Contribution of Relaxin to the Changes in Systemic Hemodynamics and Arterial Mechanical Properties during Pregnancy

A critical role for relaxin in the changes of systemic hemodynamics and arterial mechanical properties during midterm pregnancy in conscious rats was identified.[10] By administering relaxin neutralizing antibodies, the gestational increases in cardiac output and global arterial compliance, as well as the fall in systemic vascular resistance were abolished. Unexpectedly, no increase in passive compliance of isolated small renal or mesenteric arteries isolated from the midterm pregnant rats was observed. (But, this finding did corroborate earlier work.[11]) Nor was there any influence of the irrelevant control or relaxin neutralizing antibodies on arterial passive mechanics. Thus, the elevation in global arterial compliance at midgestation is a consequence of reduced vascular smooth muscle tone (i.e., vasodilation), increased number or branching of arteries, or increased passive compliance of arteries other than those investigated. Whether immuno-neutralization of circulating relaxin has complete or partial inhibitory effects during late pregnancy is presently under investigation. Possibly other hormones which are secreted by the placenta rather than the ovary contribute to systemic vasodilation (and osmoregulatory changes) during late gestation. Whether relaxin contributes to the changes in systemic hemodynamics and arterial mechanical properties during human pregnancy is currently unknown.

Because relaxin participates in both the relaxation and remodeling of the vascular wall (supra vide), there is overlap of hormonal and cellular signaling mechanisms for vasodilatory and arterial compliance changes. This sharing of hormonal and molecular mechanisms ensures a temporal coordination of the decline in both steady and pulsatile systemic arterial loads which, as previously mentioned, is critical to the maintenance of cardiovascular homeostasis during pregnancy.

During midterm pregnancy in the rat, relaxin contributes to arterial compliance changes by vasodilation, while in late pregnancy, relaxin may further increase arterial compliance by remodeling the vascular wall.[12,13]

Influence of Relaxin on the Renal Circulation

Renal Circulation during Pregnancy

In several serial studies throughout pregnancy in women, renal plasma flow (RPF) and glomerular filtration rate (GFR) were assessed by the renal clearances of para-aminohippurate and inulin, respectively (the "gold standards" for measurement of renal function, see reviews 14-16 and citations therein). On balance, these studies showed an increase in GFR and RPF of 40-65% and 50-85%, respectively, during the first half of gestation compared to prepregnant or postpartum values. Because the elevation in RPF exceeded that of GFR, the renal filtration fraction declined. During late gestation, a modest decline in RPF towards nonpregnant levels was reported, while GFR was maintained.The 24-h endogenous creatinine renal clearance is a reliable measurement of GFR in the setting of pregnancy (reviewed in ref. 14). Using this technique, Davison and Noble[17] showed that GFR rises 25% by the fourth gestational week (postLMP) compared to week 1 and reaches a 45% increase by week 9. Thus, the changes in renal hemodynamics are among the earliest and most marked maternal adaptations to pregnancy.

Animal models have been used to explore the underlying mechanisms. The Munich-Wistar rat has glomeruli belonging to the superficial cortical nephrons on the surface of the kidney that are accessible to renal micropuncture. Using this technical approach, Baylis demonstrated that the gestational rise in single nephron glomerular filtration rate was secondary to an increase in glomerular plasma flow with no change in glomerular hydrostatic pressure.[18] This constellation of events occurred as a consequence of parallel and similar decreases in the afferent and efferent renal arteriolar resistances. Although such an invasive approach cannot be used in humans, indirect evidence suggested comparable mechanisms for the gestational elevation in GFR.[19] Finally, because the gestational changes in renal hemodynamics and GFR in the chronically instrumented, conscious rat are comparable to human pregnancy, this animal model has been extensively investigated, in order to identify the underlying hormonal and molecular mechanisms.[20]

Relaxin was evaluated as a candidate hormone for the alterations in renal circulation, as well as osmoregulation during pregnancy. In pregnant rats and women circulating relaxin emanates from the corpus luteum of the ovary (reviewed in ref. 21). Human chorionic gonadotrophin (hCG) is an important stimulus for relaxin secretion during pregnancy in women.[21] In parallel with hCG, serum relaxin concentrations are highest during the first trimester (~1 ng/ml) and fall to lower levels in late gestation (~0.4 ng/ml;).[21] In fact, there were several compelling, albeit mainly circumstantial reasons to consider relaxin as the pivotal hormonal signal for triggering the circulatory and osmoregulatory changes of pregnancy as recently reviewed.[14,15]

Although renal vasodilation and hyperfiltration are detectable in gravid rats by gestational day 5, when serum relaxin is undetectable, there is a notable increase of renal function between gestational days 8 and 12 that coincides with a rise in ovarian and serum relaxin levels.[20,21] Thus, the increases in GFR and RPF that transpire in rat gestation before gestational 8 or during pseudo-pregnancy when circulating relaxin is undetectable may be mediated by other, as of yet, undetermined mechanisms. Alternatively, circulating relaxin concentrations below the level of assay detection may contribute.

Influence of Relaxin Administration on the Renal Circulation in the Nonpregnant Females and Males

Chronic, subcutaneous infusion of porcine relaxin or of rhRLX to chronically instrumented, conscious female rats for 2-5 days increased RPF and GFR (and reduced serum osmolality) to levels comparable with midterm pregnancy when renal function peaks in this species.[20,22] The renal vasodilatory response to relaxin was independent of the ovaries[22] and was also observed in male rats.[23] Chronic administration of relaxin also attenuated the renal vasoconstrictor response to an acute angiotensin II infusion[22]—a phenomenon also observed during rat pregnancy.[24-26] Moreover, the myogenic reactivity of small renal arteries isolated from the relaxin-infused rats was inhibited[27] and similar to the inhibition previously shown in small renal arteries harvested from midterm pregnant rats.[11] In contrast to the lack of effect on systemic hemodynamics and arterial mechanical properties in normotensive rats as discussed above (Influence of relaxin administration on systemic hemodynamics and arterial mechanical properties in nonpregnant females and males*)*, short-term administration of rhRLX to conscious rats for 1-4 hours produced significant renal vasodilation and hyperfiltration,[28] In normal human volunteers, short-term intravenous infusion of rhRLX for 6 hours increased RPF by 60%, but unexpectedly, not GFR.[29] The renal vasodilatory response was similar in men and women and transpired as soon as 30 minutes after starting the infusion of rhRLX. There were no significant decrements in blood pressure or serum osmolality. After 26 weeks of subcutaneous rhRLX infusion in patients with mild scleroderma, the predicted creatinine clearance rose by 15-20% and serum osmolality and blood pressure fell slightly, but significantly throughout the study in a dose dependent fashion.[30]

Contribution of Relaxin to the Changes in the Renal Circulation during Pregnancy

By administering relaxin neutralizing antibodies or removing circulating relaxin by ovariectomy while maintaining the pregnancy with exogenous estrogen and progesterone, renal hyperfiltration, vasodilation and reduced myogenic reactivity of small renal arteries were abolished in midterm pregnant rats.[5] The osmoregulatory adaptations to pregnancy were also inhibited.[5] Thus, relaxin is essential for the renal circulatory and osmoregulatory alterations during midterm pregnancy in rats.

In women who lacked ovarian function and became pregnant through egg donation, IVF and embryo transfer, the gestational rise in GFR and fall in serum osmolality were significantly subdued.[31] Because these women lacked functioning ovaries, serum relaxin was undetectable. Thus, comparable to gravid rats, endogenous circulating relaxin most likely plays a role in the initiation of the renal and osmoregulatory responses to pregnancy in women. However, unlike the gravid rat, partial responses may persist despite the absence of detectable circulating relaxin.

Cellular Mechanisms of Vasodilation

An emerging view is that the vasodilatory mechanisms of relaxin vary according to the time of exposure to the hormone, i.e., there are slow and fast vasodilatory responses.

Slow Responses (Fig. 1)

As reviewed recently, nitric oxide (NO) mediates renal vasodilation and hyperfiltration in the gravid rat model.[14-16] In contrast, the vasodilatory prostaglandins play little or no role.[14-16] In addition, NO mediates the loss of myogenic reactivity in small renal arteries obtained from midterm pregnant rats.[11] This NO dependency was also found for the renal vasodilation and hyperfiltration and loss of myogenic reactivity in small renal arteries caused by chronic relaxin administration to nonpregnant rats.[22,27] The mechanism for NO-dependent changes in the renal circulation of midterm pregnant rats or of relaxin-infused nonpregnant rats is not due to increased expression of endothelial nitric oxide synthase.[32,33]

During midterm pregnancy or during chronic subcutaneous infusion of rhRLX to nonpregnant rats, *endothelin* (ET) mediates the renal vasodilatory changes through stimulation of the endothelial ET_B receptor subtype which is linked to NO production (reviewed in refs. 14-16). Whether the expression of endothelial ET_B receptors increases, thereby constituting a primary event that mediates renal vasodilation, hyperfiltration and reduced myogenic artery of small renal arteries during pregnancy or after chronic administration of rhRLX to nonpregnant rats is presently controversial (refs. 34,35 and see below Influence of relaxin on cultured vascular cells; *The endothelial ET_B receptor*).

Typically, the traditional endothelin converting enzymes that are antagonized by phosphoramidon mediate the processing of big ET, an inactive precursor, to $ET_{1\text{-}21}$. However, phosphoramidon

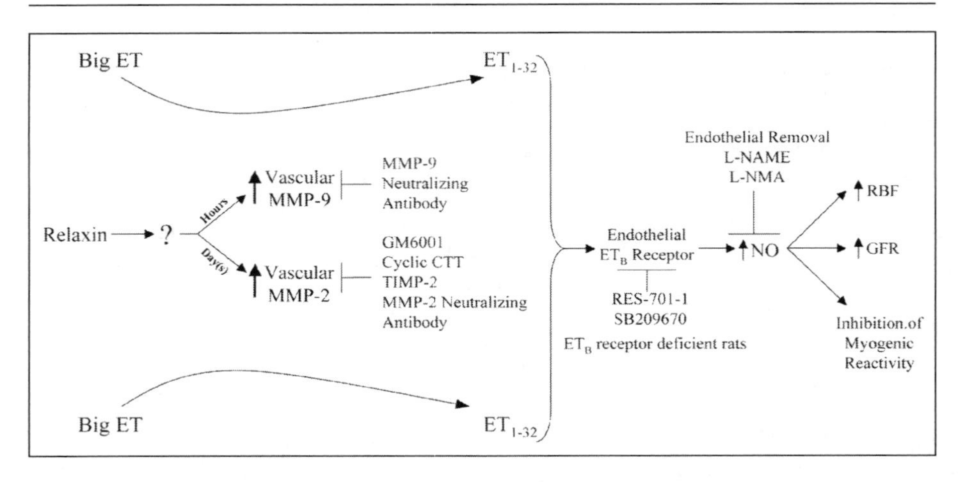

Figure 1. Working model for the slow vasodilatory actions of relaxin. ?, indicates the likelihood of another intermediary molecule. ┤, indicates inhibitors of relaxin vasodilation. ET, endothelin; MMP, matrix metalloproteinase; RBF, renal blood flow; GFR, glomerular filtration rate; GM6001, a general MMP inhibitor; cyclic CTT, a specific peptide inhibitor of MMP-2; TIMP-2, tissue inhibitor of metalloproteinase; RES-701-1, a specific ET_B receptor antagonist; SB209670, a mixed ET_A and ET_B receptor antagonist; L-NAME, nitro-L-arginine methyl ester; L-NMA, N^G-monomethyl-L-arginine. Note that phosphoramidon (an inhibitor of the classical endothelin converting enzyme), STT (control peptide for cyclic CTT); heat inactivated TIMP-2, BQ-123 (a specific ET_A receptor antagonist) and IgGs (control antibodies for MMP neutralizing antibodies) did not affect the slow vasodilatory actions of relaxin. See text for details and references.

did not affect renal vasodilation, hyperfiltration or reduced myogenic reactivity of small renal arteries harvested from nonpregnant rats chronically administered rhRLX.[36] Therefore, based on the confluence of our own and others experimental findings, we hypothesized that relaxin may increase vascular *matrix metalloproteinase-2* (MMP-2) activity in renal arteries during pregnancy that, in turn, cleaves big ET to ET_{1-32} at a glycine-leucine bond, thereby accentuating the endothelial ET_B receptor-NO vasodilatory pathway.[36]

After acute infusion of a specific inhibitor of gelatinase (MMP-2 or -9) or of a general inhibitor of MMP to conscious rats during chronic administration of rhRLX, renal vasodilation and hyperfiltration were abrogated. Furthermore, the absent myogenic reactivity of small renal arteries obtained from midterm pregnant rats or from nonpregnant rats chronically administered rhRLX was restored by incubating the arteries with the specific MMP-2 or -9 inhibitor, the general MMP inhibitor, TIMP-2 (tissue inhibitor of metalloproteinase), or a specific MMP-2 neutralizing antibody, but not by phosphoramidon. In addition, MMP-2 activity was augmented in small renal (and mesenteric) arteries isolated from midterm pregnant rats or nonpregnant rats chronically administered the hormone. An elevation in MMP-2 activity in response to chronic subcutaneous rhRLX infusion was also seen in small renal arteries from rats that were genetically deficient in the ET_B receptor. However, myogenic reactivity remained robust and was not the least bit attenuated in these arteries. The latter finding corroborated the essential role of the endothelial ET_B receptor in mediating the inhibition of myogenic reactivity by rhRLX as previously determined using pharmacological inhibitors of the receptor.[14-16] Similar results were observed in midterm pregnant rats deficient in the ET_B receptor.[37] Of greater significance, however, is the dissociation of increased vascular MMP-2 activity from the inhibition of myogenic reactivity strongly suggests that MMP-2 is in series with and upstream of, the endothelial ET_B receptor-NO vasodilatory pathway).

The mechanism(s) underlying the increase in vascular MMP-2 activity by chronic rhRLX administration or pregnancy is incompletely understood. However, both pro and active MMP-2 activities are elevated by a similar degree, MMP-2 protein and mRNA are also increased and there are no significant alterations in TIMP-1 and 2.[38] MMP-2 protein is observed in both endothelium and smooth muscle, but further investigation is required to determine in which of these vascular compartment(s) it increases in response to pregnancy or chronic rhRLX administration. Although it is likely that the relaxin LGR7 receptor mediates the arterial responses to pregnancy and chronic relaxin administration in nonpregnant rats, this supposition needs testing. Furthermore, whether the relaxin receptor which mediates the slow vasodilatory responses is located in the endothelium or smooth muscle and whether there are intermediary molecules linking relaxin receptor activation to increased MMP-2 expression are presently unknown.

Recently, we reported that *matrix metalloproteinase-9* (MMP-9) rather than MMP-2 activity is elevated in small renal and mesenteric arteries isolated from rats after short-term subcutaneous infusion of rhRLX for 4-6 hours.[39] Small renal arteries demonstrated loss of myogenic reactivity and robust myogenic reactivity was restored by incubation with a specific MMP-9 antibody, rather than a specific MMP-2 antibody as observed following day(s) of rhRlx administration (supra vide). Like MMP-2, MMP-9 can process big ET at a glyine-leucine bond. The inhibition of myogenic reactivity after 4-6 hours of rhRLX administration was also mediated by the endothelial ET_B receptor-nitric oxide vasodilatory pathway. Preliminary studies suggest that these slow vasodilatory responses after hours to day(s) of hormone exposure require cross-talk between the endothelium and vascular smooth muscle involving vascular endothelial growth factor (VEGF) in one direction (vascular smooth muscle to endothelium), in addition to nitric oxide in the other direction (endothelium to vascular smooth muscle).[40]

Fast Responses

Arteries isolated from human gluteal biopsies and mounted in a wire myograph, relaxed to rhRLX after preconstriction with norepinephrine in an endothelium-dependent fashion. However, human pulmonary resistance arteries did not vasodilate in response to rhRlX.[41] Human subcutaneous arteries studied in a pressure arteriograph and preconstricted with phenylephrine demonstrated a rapid and marked vasorelaxation to rhRLX that was mediated by phosphatidylinositol 3-kinase and NO.[42] Similar findings were observed for rat small renal, but not mesenteric or coronary (septal) arteries.[42] Thus, the ability of relaxin to produce rapid relaxation in isolated human and rat blood vessels is apparently dependent on their anatomical source.

Influence of Relaxin on Cultured Vascular Cells

Inducible Nitric Oxide Synthase

In cultured vascular smooth muscle cells from bovine aorta, Bani et al provided evidence that purified porcine RLX induced iNOS expression.[43] The porcine relaxin was stated to be endotoxin-free as determined by the Limulus amebocyte lysate assay. They reported that porcine RLX stimulated calcium and calmodulin-independent NOS activity, upregulated iNOS protein by immunohistochemistry, elevated nitrite in the conditioned media, increased cGMP in the cultured cells and blocked the rise in the calcium transient by thrombin. The same investigators utilized similar methodological approaches to show that porcine RLX could upregulate iNOS in cultured rat coronary artery endothelial cells[44] and in cultured human umbilical artery endothelial cells.[45] Whether RLX induces iNOS in vivo was not investigated, although the authors presented data supporting iNOS induction by RLX in hearts challenged with anaphylaxis (normal hearts were apparently not investigated, see below Relaxin and cardiac disease; *Cardiac anaphylaxsis*).

It should be pointed out that not all evidence supports the concept that relaxin induces iNOS expression insofar as (i) infusion of porcine RLX to conscious rats failed to increase the urinary excretion of cGMP and of nitrate plus nitrite,[22] (ii) calcium-independent NOS activity was not increased in isolated mesenteric arcades or thoracic aortae from rats of gestational days 15-17 when circulating levels of relaxin are high[46] and (iii) iNOS expression was not induced in renal tissues by pregnancy (or rhRLX) as evaluated by Western analysis at least in one study,[32] although it may have been in another.[33]

The Endothelial Endothelin B Receptor

Dschietzig and coworkers reported that human RLX augmented the expression of the ET_B receptor subtype in cultured human umbilical artery endothelial cells.[35] The hormone elevated ET_B receptor mRNA, protein and binding sites for radiolabeled ET-1. RLX activated extracellular signal-regulated kinases-1/2 (ERK-1/2), but not p38 kinase and enhanced the DNA binding activity of the transcription factor, nuclear factor NF-κB; pharmacological blockade of these signal transduction pathways prevented the elevation in endothelial ET_B receptor mRNA expression. Many of these findings were duplicated in cultured bovine aortic endothelial cells. The expression of the ET_B receptor by human vascular smooth muscle cells was unaffected by RLX, nor was the ET_A receptor subtype altered in any of the cultured cells by the hormone. Relaxin also increased NF-κB reporter activity in bovine aortic endothelial cells which was not affected by dominant negative Ras (a GTP binding protein coupled to growth factor receptors), but was inhibited by dominant negative forms of mitogen activate protein kinase kinase-1 (MEK-1), Raf-1 (a MEK kinase) and ERK-1/2, thereby providing further clues to the cell signaling mechanisms. The functional importance of these findings was shown in arteries from rats, insofar as incubation with RLX in vitro resulted in a decrease in the sensitivity and maximal response to ET-1-induced contraction. Furthermore, the vasorelaxation to ET-3 was enhanced. All of these functional effects were blocked by the ET_B receptor antagonist, A-192621 or by the MEK-1 inhibitor, PD-98059. On balance, these results corroborate those obtained from the renal circulation where pregnancy or relaxin-treatment of nonpregnant rats elicited renal vasodilation, hyperfiltration and impaired myogenic reactivity of small renal arteries via the endothelial ET_B receptor/NO vasodilatory pathway.[5,11,23-25,27,36,47] The study of Dschietzig and colleagues further suggests that one potential mechanism for this vasodilatory role of RLX is through upregulation of the endothelial ET_B receptor.

The concept that relaxin upregulates the endothelial ET_B receptor is both logical and attractive and Kerchner et al also tested the hypothesis. Unfortunately, they were unable to find any supportive evidence.[34] Briefly, ET_B expression was investigated in small renal arteries that were harvested from virgin and midterm pregnant rats, as well as from nonpregnant rats that were infused subcutaneously with recombinant human relaxin (rhRLX) or vehicle for 5 days or for 4-6 hours. Small renal arteries harvested from additional virgin rats were incubated with rhRLX or vehicle for 3 hours at 37°C in vitro. ET_B expression was also assessed in cultured human endothelial cells: aortic (HAEC), coronary (HCAEC), umbilical vein (HUVEC) and dermal microvascular endothelial cells (HMVEC). Cells were incubated for 4, 8, or 24 hours with rhRLX (5, 1, or 0.1ng/ml) or with vehicle. ET_B protein in arteries and cells was evaluated by Western analysis. No alteration of ET_B expression was observed in small renal arteries from any of the experimental protocols. Nor was there an augmentation of the vasorelaxation response to ET-3 in small renal arteries incubated in vitro with rhRLX. rhRLX only sporadically altered ET_B expression in HCAEC and HUVEC at certain time points or doses and no regulation was observed in HAEC or HMVEC. These results suggest that up-regulation of ET_B receptor protein has little or no role in relaxin stimulation of the endothelial ET_B/NO vasodilatory pathway. Rather, the pivotal step regulated by relaxin is vascular gelatinase activity (see above Cellular mechanisms of vasodilation; *Slow responses*).

Influence of Relaxin on the Circulations of Other Organs

Coronary Blood Flow

Bani-Sacchi and colleagues demonstrated that relaxin was a potent coronary vasodilator.[48] Hearts were harvested from male guinea pigs and rats, mounted in a Langendorff apparatus and perfused retrograde through the aorta at constant pressure. The coronary effluents were collected for determination of coronary flow and nitrite (a metabolite of NO). Whether administered by bolus injection into the aortic cannula (1, 5 and 10 nM) or by addition to the perfusion media for constant infusion (5 nM), there was a fast and parallel increase in coronary blood flow and nitrite concentration. Both were prevented by pretreatment with 0.1 mM L-NMMA, an inhibitor of NO synthase. In these protocols, relaxin was found to be more potent than acetylcholine (plus physostigmine to inhibit acetylcholinesterase) and sodium nitroprusside, endothelium-dependent and –independent vasodilators, respectively.

Uterus

The first evidence suggesting that relaxin affects blood vessels was based on histological studies conducted in the uterus. In 1966, Dallenbach-Hellweg et al reported an enlargement of arterioles and capillaries in the superficial part of the endometrium of castrated monkeys treated with RLX.[1] In a later study also in monkeys, they described not only an enlargement of the vessels but also endothelial cell proliferation[2] Vasilenko and colleagues published similar findings in ovariectomized rats administered porcine RLX.[49] Finally, immunohistochemical studies demonstrated binding sites for RLX on cells associated with blood vessels in the uterus, cervix and vagina of pigs and humans.[50,51] These studies are consistent with a vasodilatory role for RLX in the uterine circulation.

The results of functional studies, however, are less clear. rhRLX treatment has been shown to increase uterine blood flow in conscious ovariectomized female rats.[52] In contrast, porcine RLX had no effect on endometrial and myometrial blood flow in mature anesthetized sheep.[53] It is likely that species differences explain this discrepancy. Sheep do not produce RLX because the mRNA contains numerous stop codons in the C-peptide region which prevent the translation of a functional RLX molecule.[54] To our knowledge, there are no studies directly examining the effects of RLX on uterine blood flow in humans. However, in phase II/III trial of rhRLX in the treatment of scleroderma, the most frequently reported adverse event by women receiving the hormone was heavy or irregular menstrual bleeding suggesting increased endometrial vascularization. In addition, rhRLX increases expression of VEGF in human endometrial cells in culture.[55] These findings support a role for RLX in the regulation of human endometrial blood flow. In contrast, a study in pregnant women showed a positive correlation between plasma RLX levels and the resistance or pulsatility index measured by ultrasound early in gestation. This finding suggests that RLX may increase uterine artery resistance during pregnancy.[56] Finally, in an in vitro study of human intramyometrial arteries, rhRLX had no effect on either resting tension or tension induced by U46619 (a thromboxane analog), endothelin or PGF_{2a}[57]. A role for RLX in the modulation of uterine blood flow requires further investigation.

Placenta

RLX binding sites were found on blood vessels in the chorionic plate and within the placental villi.[50] Despite the fact that binding sites for RLX were identified, rhRLX did not affect resting tension or tension induced by U46619, endothelin-1 or $PGF_{2\alpha}$ in human placental stem villous arteries studied in vitro.[57] Another study involving human umbilical arteries failed to demonstrate an effect of RLX or RLX and progesterone on serotonin and KCl stimulated contractions in vitro.[58]

Mammary Gland

Evidence that RLX may act as a vasodilator in the mammary circulation includes morphometric analyses of microvessel lumina from ovariectomized mice 18-20 hours after a single injection of porcine RLX. In this study, the mean diameters of arterioles, capillaries and postcapillary venules in the mammary glands of the RLX-treated mice were significantly greater when compared to ovariectomized control mice.[59] It is possible that this effect of RLX is not limited to the mouse since RLX binding sites were identified in blood vessels of the human mammary gland and nipple.[50,51] The pigeon crop sac is a structure analogous to the mammary gland and there are several advantages to studying this organ. One is that hormones can be administered by intradermal injection which permits study of local effects. Another advantage is that the anatomy of the crop sac allows for treatment of one hemi-crop, while the contralateral crop serves as control. In this tissue, there was striking dilation of the blood vessels in the lamina propria of mucosa 6 hours after injection of porcine RLX in one hemi-crop compared to the vehicle-treated contralateral crop.[60]

Liver, Mesentary and Mesocaecum

The influence of RLX on the hepatic vasculature was investigated in male rats. In this study, porcine RLX dilated the liver sinusoids. This effect was reduced by NO synthesis inhibition.[61]

In the mesenteric circulation, RLX modulated the vascular responses to vasoconstrictor agents. In the perfused mesentery of the spontaneously hypertensive rat in situ, the vasoconstrictor responses to arginine vasopressin and norepinephrine were attenuated after a 42 hour infusion of purified rat RLX.[62] In the same investigation, the sensitivity to norepinephrine was also reduced in the isolated portal vein from RLX-treated rats while the sensitivity to angiotensin was unchanged. After a six hour incubation of isolated mesenteric and renal arteries, as well as aorta from rats with rhRLX, the maximal contractile response and sensitivity to ET-1 decreased while the relaxation response to ET-3 was augmented.[35] (ET-1 contracts arteries by interacting with ET_A and/ET_B receptor subtypes on vascular smooth muscle. ET-3 relaxes arteries by interacting with the ET_B receptor subtype on endothelium.) These functional responses suggested that RLX upregulates the endothelial ET_B receptor, thereby offsetting the contractile response to ET-1 and potentiating the relaxation response to ET-3 (also see above Influence of relaxin on cultured vascular cells; *The endothelial ET_B receptor*). Mesenteric arteries isolated from rats after chronic administration of rhRLX were also less responsive to changes in intraluminal pressure (reduced myogenic reactivity) compared to vehicle treated rats when studied in a pressure arteriograph system.[27]

In the mesocaecum, administration of porcine RLX to male Wistar rats elicited a rapid dose-dependent dilation of the veins; however, arteriolar and capillary flows were unchanged.[63] In the same study, RLX administration also opposed the vasospasm induced by norepinephrine or promethazine (an anticholinergic agent) in the arteries of the mesocaecum.

Relaxin and Angiogenesis

Vasodilatory hormones frequently have an angiogenic role (or vice versa). One of the many interesting observations to arise from the phase II/III trial of rhRLX in the treatment of scleroderma was the high incidence of menometrorrhagia (heavy, irregular or prolonged menstrual bleeding) in those women administered hormone.[55] In cultured human endometrial cells, rhRLX stimulated the production of VEGF in a cAMP-dependent fashion.[55] rhRLX stimulated new blood vessel formation in the endometrium of a nonhuman primate model of early pregnancy.[64] Others have also reported the angiogenic properties of relaxin in various settings.[65-67] Taken together, these results suggest that relaxin contributes to neovascularization.

Unemori and colleagues reported enhanced vessel in-growth of Matrigel plugs containing rhRLX that were implanted subcutaneously in mice.[68] Using the Hunt-Schilling wound chamber assay, they documented increased VEGF 164 and basic fibroblast growth factor mRNA expression by cells (presumably macrophages) in the chamber aspirate of those rats administered rhRLX systemically. This increase corresponded with enhanced expression of factor VIII related antigen in the granulation tissue of the chamber, indicative of enhanced blood vessel growth. Interestingly, VEGF mRNA expression was not elevated by rhRLX in resident macrophages of the lung or spleen remote from the site of injury. Nor did rhRLX have a direct angiogenic effect on cultured endothelial cells confirming an earlier report.[69] Finally, rhRLX was also shown to increase VEGF 165 and 120, as well as bFGF mRNA expression by THP-1 cells, a human monocyte/macrophage cell line. Taken together, rhRLX appears to be indirectly angiogenic by enhancing VEGF and bFGF expression in wound macrophages, thereby stimulating angiogenesis. In a subsequent investigation by the same investigators, systemic administration of rhRLX was demonstrated to significantly accelerate wound closure in full thickness skin excisions in db/db mice.[70] This accelerated healing was associated with increased staining for factor VIII related antigen in the granulation tissue which was thicker and more cellular than in vehicle-treated mice.

Further evidence for the indirect angiogenic attributes of rhRLX was shown in a model of chronic myocardial infarction in rats.[71] Systemic infusion of rhRLX potentiated the increase in bFGF mRNA and protein at 7 and 21 days in the peri-infarct region where the growth factor was expressed by myocytes and fibroblasts. rhRLX infusions in sham operated rats showed no change in bFGF or VEGF expression in corresponding regions of the left ventricle or in the right ventricle. The increase in the number of thin-walled, collateral vessels lacking smooth muscle was correspondingly potentiated in the peri-infarct region by systemic rhRLX. In the same publication, rhRLX was reported to increase the mRNA and protein for VEGF and bFGF in human neonatal heart cells in culture (fibroblasts and myocytes).

Evidence for a Vascular-Derived Relaxin Ligand-Receptor System

We explored whether there is local expression and function of relaxin and its receptor in arteries of nonpregnant females and males.[72] Relaxin-1 and its major receptor, LGR7, mRNA were expressed in thoracic aortae, small renal and mesenteric arteries from mice and rats of both sexes, as well as in small renal arteries from female tammar wallabies (an Australian marsupial). Using antibodies available for rat and mouse LGR7 receptor and rat relaxin, we also identified protein expression in arteries. These results extended the findings of Dscheitzig et al who showed H1 and H2 relaxin mRNA expression in human saphenous vein and mammary artery,[73] as well as those of Sherwood and colleagues who demonstrated relaxin binding sites on blood vessels in reproductive organs.[50,51] Small renal arteries harvested from relaxin-1 gene deficient mice demonstrated enhanced myogenic reactivity and decreased passive compliance relative to wild-type and heterozygous mice.[72] Taken together, these findings suggest an arterial-derived, relaxin ligand-receptor system that acts locally to regulate arterial function.

Part 2. Relaxin and Vascular Dysfunction: Implications for Relaxin as a Therapeutic Agent (Table 2)

Relaxin and Renal Disease

Given the renal vasodilatory (see above Influence of relaxin on the renal circulation) and matrix-degrading[74-76] properties of relaxin, there are important therapeutic implications of this hormone in disorders associated with renal vasoconstriction and fibrosis. While scarring and fibrosis are late findings of renal disease, renal alterations may occur prior to clinically recognized renal insufficiency. At this time period, the hemodynamic effects of relaxin may also have therapeutic benefit. As described above, relaxin administration, similar to pregnancy, results in renal vasodilation and hyperfiltration in rats via an endothelial endothelin B and nitric oxide pathway in rats. Renal micropuncture in pregnant Munich-Wistar rats by Baylis and colleagues indicates that the vasodilation of the efferent and afferent arterioles occurs without a net increase in glomerular capillary pressure.[18] Thus, relaxin may have therapeutic implications in renal disorders characterized by increased intraglomerular pressure secondary to efferent arteriolar constriction. Smith and colleagues administered intravenous rhRLX to healthy male and female volunteers over 6 hours and noted a 47% increase in renal plasma flow, but no significant change in glomerular filtration rate.[29] There were no adverse effects suggesting that relaxin may be used in humans and could have potential therapeutic benefit in increasing renal blood flow in certain disease states. Longer infusions may also increase GFR.[22,30]

Despite varying etiologies of progressive renal diseases, scarring is the final common pathway leading to renal insufficiency, end stage renal disease, dialysis and need for transplantation. Fibrosis can affect both the renal parenchyma and vascular components of the kidney leading to tubulointerstitial scarring and glomerulosclerosis, respectively. The finding that relaxin-deficient mice (RLX -/-) develop an age-related progressive fibrosis in the kidney and other tissues[77] underscores the role of relaxin in the regulation of matrix turnover and fibrosis. Furthermore, exogenous administration of relaxin in several rodent models of renal disease results in improvement; these include bromoethylamine-induced renal interstitial fibrosis,[78] ablative and infarction models of 5/6 nephrectomy,[79] nephrotoxicity related to cyclosporine,[80] anti-glomerular basement membrane nephritis (Goodpasture's syndrome),[81] and RLX -/- renal fibrosis.[82] Danielson and colleagues demonstrated that rhRLX administration to aged Munich Wistar rats improved both renal function and histology.[83] GFR and ERPF were increased and effective renal vascular resistance decreased with long-term RLX-infusion over days. Double-blinded histologic examination of the kidneys revealed a decrease in glomerular and tubular collagen deposition in the RLX-treated compared to vehicle-treated controls. With short-term (24 hours) RLX-administration, the improved renal function was mediated by vascular gelatinase activity. This was demonstrated by infusion of a specific gelatinase inhibitor, cyclic CTT and a general MMP inhibitor, GM6001, as well as upregulation of vascular MMP-2 activity on gelatin zymography (see above Cellular mechanisms of vasodilation).

Table 2. ***Relaxin and vascular dysfunction: potential therapeutic implications***

Target Organ System	Specific Disease Process	Therapeutic Benefi
Kidneys	Renal vascular disease—vasodilation	• Vasodilation of pre- and post-glomerular arterioles without increase in intraglomerular pressure in rats (18) • Increased renal blood flow in healthy male and female volunteers with IV RLX (29) • Increased predicted creatinine clearance in patients with mild scleroderma during chronic subcutaneous RLX administration (30)
	Renal fibrosis	• Age-related renal fibrosis in RLX-deficient mice(77) • Bromoethylamine-induced renal fibrosis (78), models of 5/6 nephrectomy (79), cyclosporine nephrotoxicity (80), anti-glomerular basement membrane nephritis (81) • Renal histology and function in aged rats (83)
Heart	Myocardial ischemia-reperfusion (I/R) injury	• I/R with coronary artery ligation in guinea pig, rat and swine models (85, 86,93,94) • Isoproteronol-induced myocardial ischemia in rats (92)
	Cardiac fibrosis	• Age-related cardiac fibrosis in male RLX -/- mice (95) • β2 adrenergic receptor overexpressing transgenic mice (96) • Spontaneously hypertensive rats (99,100)
	Heart failure	• Afterload reduction (6,8-10)
	Post-infarction heart	• Improved integration of transplanted cells in permanently damaged heart – in vitro coculture model (94,106)
	Cardiac anaphylaxis	• Protection against cardiac anaphylaxis in ex vivo guinea pig heart model (107)
Brain	Cerebral ischemia (stroke)	• Rodent model of stroke (middle cerebral artery occlusion) (108)
Gastrointestinal tract	Intestinal ischemia-reperfusion	• Rat model with splanchnic artery occlusion (111)
	Liver transplantation—preservation and reperfusion	• Isolated perfused rat liver model (112)
Pulmonary	Pulmonary hypertension	• Hypoxia model of pulmonary hypertension in rats (113)
Blood vessels	Vascular inflammation	• In vitro co-culture model of coronary artery endothelial cells and neutrophils (88)
Pregnancy	Preeclampsia	• Potential benefit based on angiogenic effects, renal and systemic vasodilation in rats and humans (above) and uterine artery vasodilation in rats (52)

Interestingly, after 20 days, the improved renal function was insensitive to gelatinase inhibition, suggesting a permanent alteration in the vascular structure.

The anti-fibrotic mechanisms of relaxin and potential therapeutic benefits of relaxin are discussed further in the chapter by Dr. Samuel and coworkers.

Relaxin and Cardiac Disease

In addition to systemic and renal vasodilatory effects, relaxin has direct cardiac effects that may be important in both health and disease. Relaxin is produced by the heart[73,84] and acts on specific cardiac receptors. There are also direct positive chronotropic and inotropic effects on the heart (at least in rodents), as well as platelet inhibitory effects of relaxin that may play a role in cardiovascular health and disease.

Myocardial Ischemia-Reperfusion Injury

Masini and colleagues demonstrated that administration of exogenous relaxin improves myocardial injury secondary to ischemia-reperfusion in a rodent model.[85] Using isolated hearts from male guinea pigs, myocardial ischemia was produced by ligating the left anterior descending coronary artery. When porcine relaxin (30 ng/ml) was added to the perfusate at the start of coronary occlusion, coronary flow ligation was preserved despite ligation most likely secondary to dilation of the extensive collateral circulation in the guinea pig heart. In addition, RLX augmented coronary flow during reperfusion. Furthermore, nitrite concentration was increased in the coronary effluent, histamine release from mast cells was reduced, markers of oxygen free-radical mediated cardiomyocyte injury including malondialdehyde and calcium content were also reduced in the relaxin-treated hearts.

Studies of the beneficial effects of relaxin in cardiac ischemia-reperfusion injury were extended to an in vivo rat model by Bani and colleagues.[86] Thoracotomy was performed on anesthetized rats and the left anterior descending coronary artery was ligated for 30 minutes followed by 60 minutes of reperfusion. Rats administered porcine RLX (100 ng intravenously) 30 minutes prior to occlusion had significantly reduced area of myocardial damage, fewer ventricular arrhythmias and reduced mortality. In addition, neutrophil accumulation, lipid peroxidation, calcium content, endothelium and myocyte injury, mast cell degranulation and histamine release were all reduced with relaxin pretreatment. This group of investigators has also posited that relaxin may have direct effects on neutrophils to prevent activation and adhesion to endothelial cells,[87,88] on mast cells to inhibit degranulation and histamine release,[89,90] on myocytes to reduce calcium metabolism and on platelets to reduce aggregation[91] possibly via a nitric oxide-mediated mechanism.

Using a rat model of isoproterenol-induced myocardial ischemia, Zhang and colleagues[92] demonstrated increased rat relaxin-3 protein and mRNA in the myocardium and increased plasma relaxin-3 (the latter measured by RIA from Phoenix Pharmaceutical Inc.). Furthermore, administration of exogenous relaxin (0.2 and 2.0 μg/kg/day) improved cardiac function, reduced levels of malondialdehyde, lactate dehydrogenase and creatine phosphokinase in the plasma of relaxin-treated rats compared to controls. With the higher dose of relaxin, fibroblastic hyperplasia in the myocardium and circulating endothelin were reduced. These findings suggest a protective role for endogenous relaxin and a therapeutic role for exogenous relaxin in this model of myocardial injury.

More recently, the therapeutic potential of recombinant human relaxin in myocardial ischemia-reperfusion injury has been extended to a swine model[93,94] which has been used extensively to test cardiac drugs and therapies due to intrinsic similarities to human ischemic heart disease. In this approach, the investigators sought to mimic the clinical situation of acute myocardial infarction followed by admission to an intensive care setting and intervention with percutaneous coronary artery angioplasty. The current standard of care for acute myocardial infarction includes catheter-based reperfusion with either balloon angioplasty with or without coronary artery stenting. Preventing reperfusion injury with relaxin is the basic hypothesis of these most recent studies. Using the swine model, thoracotomy and catheterization was performed under anesthesia followed by occlusion of the left anterior descending artery for 30 minutes.[93] Relaxin was administered for

20 minutes starting at reperfusion. Three doses of RLX, 1.25, 2.5 and 5 μg/kg bodyweight, were utilized. At the two higher doses, there was reduction of the main serum markers of myocardial damage (myoglobin, creatine kinase—MB fraction, troponin T), reduced tissue parameters of oxygen free-radical induced cardiomyocyte injury (malodialdehyde, tissue calcium), reduced cardiomyocyte apoptosis (caspase 3) and reduced inflammatory leukocyte recruitment (myeloperoxidase) compared to vehicle-treated controls. Furthermore, there was a reduction in the tissue volume that did not uptake the tracer ^{201}Thallium and less ultrastructural evidence of cardiac damage. Cardiac contractile performance was also improved at the 5 μg/kg dose of relaxin.

Cardiac Fibrosis

A variety of causes result in cardiac fibrosis including but not limited to acute inflammation, myocardial infarction, aging, increased hemodynamic load, neurohumoral activation and metabolic disorders. Hypertensive disorders and ischemic heart disease resulting in cardiac fibrosis and/or hypertrophy are of particular clinical importance. Evidence suggests that endogenous relaxin may modulate the extracellular matrix turnover and scarring and exogenously administered relaxin may have therapeutic benefit. This topic is reviewed extensively by Samuels and colleagues elsewhere in this book, but will be summarized briefly here.

Male relaxin-deficient (RLX -/-) mice demonstrate an age-related progression of cardiac fibrosis.[95] Increased left ventricular collagen content and concentration as well as increased atrial hypertrophy, left ventricular procollagen I mRNA expression, chamber stiffness and diastolic dysfunction were noted in relaxin-deficient mice compared to wild-type controls from nine months of age onwards.[95] Interestingly, these features were not observed in relaxin-deficient female mice. Local expression of relaxin and relaxin receptors supports the idea of the cardiac tissue as a potential source and/or target for relaxin. Using RT-PCR, relaxin and relaxin-3 gene transcripts have been identified in the atria and ventricles of rodent hearts[95-97] and H1 and H2 RLX are constitutively expressed in human cardiovascular tissues and upregulated in disease states such as heart failure.[73] Binding sites have also been identified in the atrium of male and female rats.[98]

Exogenous relaxin has cardiac anti-fibrotic properties in various rodent models. Treatment of established cardiac fibrosis in 12 month old RLX -/- mice with two weeks of rhRLX reduced collagen deposition.[96] Infusion of rhRLX for 14 days reversed cardiac fibrosis in transgenic animals overexpressing β2 adrenergic receptors which eventually develop heart failure.[96] In the spontaneously hypertensive rat model, rhRLX administration to 9-10 month old rats significantly decreased collagen content in the myocardium of the left ventricle, but not the unaffected chambers of the heart.[99] Another group of investigators demonstrated left heart hypertrophy and selective elevation of left atrial and ventricular relaxin gene expression and relaxin peptide in 12-month old spontaneously hypertensive rats.[100] Work in the arena of cardiac fibrosis suggests that endogenous relaxin may have a protective role in cardiac injury and fibrosis and exogenous relaxin may have a therapeutic role in preventing or reducing fibrosis in the heart. Further investigation is needed before extrapolating these findings to human disease.

Heart Failure

Dschietzig and colleagues demonstrated an association between congestive heart failure (CHF) and circulating relaxin levels in humans.[73] Circulating RLX levels correlated with disease severity compared to controls. In 11 of 14 subjects with severe CHF, the RLX concentration in the coronary sinus was greater than the left ventricle suggesting a cardiac contribution to the elevated circulating RLX levels. Furthermore, after 12 to 48 hours of sodium nitroprusside infusion, plasma RLX levels in severe CHF patients fell to those seen in moderate CHF. There was no change in RLX with nitroprusside infusion in the moderate CHF group. RLX levels also correlated with left ventricular end diastolic pressure ($r = 0.69$) and cardiac index ($r = -0.62$). In tissues from the failing heart, these authors noted increased H1 and H2 gene expression in the right atrium and left ventricle, increased 18-kDa prorelaxin but not the 6-kDa mature form in the right atrium. Interestingly, levels of the prohormone convertase-1 mRNA were decreased in the right atrium but not the left ventricle possibly accounting for the increase in prorelaxin in the atrium but not the ventricle. It should be noted, however, that not all investigators observed an association between circulating levels of relaxin and heart failure[101] (and see Controversies, Unresolved Issues and Future Directions). Based on the unique constellation of effects in the systemic and renal circulations, relaxin may be a particularly effective therapeutic agent for afterload reduction in congestive heart failure.[6,16]

Post-Infarction Heart

A newer research area in the therapeutic potential of relaxin has been in the post-infarction heart and permanently damaged tissue, including scarring and aneurysms. There has been interest in precursor cell grafting to repair the post-infarct myocardium.[102,103] Currently, skeletal muscle myoblasts have garnered particular attention because of high tolerance to ischemia, high proliferative potential and autologous source.[103-105] On the basis of relaxin's anti-fibrotic effects, angiogenic potential and influence on cardiomyocytes in fetal and neonatal periods, Bani and colleagues have performed preliminary studies using relaxin to improve integration of the grafted cells into the host cardiac tissue.[94,106] Using in vitro coculture of mouse skeletal myoblasts and adult rat cardiomyocytes, relaxin increased gap junction formation with increased connexin 43 expression on myoblasts and potentiated gap junction-mediated intracellular exchanges including increased conductance and intercellular transmission of calcium signals. These preliminary investigations suggest a novel role for relaxin in promoting effective transplantation and differentiation of cells into the damaged heart.

Cardiac Anaphylaxis

Masini and coworkers have also shown a beneficial effect of RLX in protecting against cardiac anaphylaxis in an ex vivo guinea pig heart model.[107] Two intraperitoneal injections of ovalbumin were administered on consecutive days to male guinea pigs for the purpose of sensitization. Fifteen to 30 days later, the heart was harvested and studied ex vivo using a Langendorff apparatus. Cardiac anaphylaxis was induced by administration of ovalbumin into the aortic cannula. In order to study the effects of relaxin, porcine RLX (30 ng/ml) was administered to the perfusate 30 minutes prior to the final ovalbumin injection. RLX attenuated the reduction in coronary flow and adverse inotropic and chronotropic effects caused by the ovalbumin challenge. In addition, the increases in histamine content in coronary effluent and myocardial tissue as well as mast cell degranulation were inhibited by RLX. Increased nitrite concentration, cardiac expression of iNOS and cGMP along with reduced calcium content were observed with relaxin treatment compared to controls. Based on these findings and prior work by the same authors on bronchial hyper-responsiveness and inflammatory lung injury induced by antigen challenge,[89] the mechanism by which RLX protects against cardiac anaphylaxis is via an inhibition of mast cell degranulation and preventing release of histamine and other mediators.

Relaxin and Cerebral Ischemia (Stroke)

Based on the reported properties of relaxin in the cardiovascular system and the presence of relaxin binding sites and actions on the brain, it has been hypothesized that relaxin may have a neuroprotective effect in stroke and ischemic brain disease.[108,109] Wilson and colleagues have performed preliminary studies using a rodent model of stroke.[108] Recombinant human RLX (10ng in 200 nL of saline) or 200 nL of saline was injected into the secondary somatosensory cortex of anesthetized rats. Thirty minutes after treatment, the middle cerebral artery was occluded in two areas using bipolar coagulation. After four hours, the rats were killed and brains isolated. The ratio of the infarct area to the ipsilateral hemispheric area was compared between the relaxin-treated and control groups. Proper location of injection was confirmed by analyzing brain sections. Based on four rats in each treatment group, relaxin pretreatment significantly reduced the mean infarct area/hemispheric area ratio. The investigators speculate that the possible mechanisms of the reduced tissue death with RLX treatment could be secondary to nitric oxide mediated vasodilation thereby improving perfusion and/or improved collateral circulation or alternatively via activation of estrogenic mechanisms which are neuroprotective. Recently published work does not support the estrogenic mechanisms that were posited in this model.[110] Relaxin-induced VEGF mediated mechanisms are also postulated as being a potential contributing factor.

Relaxin and Intestinal Ischemia-Reperfusion Injury

Intestinal ischemia secondary to reduced or absent blood flow can result in endothelial injury and inflammation contributing to the pathophysiology of shock. Masini and colleagues have applied their extensive experience with cardiac ischemia-reperfusion injury to a well recognized animal model of splanchnic artery occlusion followed by reperfusion (SAO/R).[111] All rats underwent surgical splanchnic artery occlusion and the three groups were administered purified porcine RLX (30 ng/kg), inactivated RLX, or vehicle 15 minutes prior to reperfusion. In addition to reducing the drop in blood pressure caused by SAO/R, bioactive relaxin also reduced leukocyte infiltration as measured by myeloperoxidase activity and expression of endothelial cell adhesion markers in the ileum. RLX also reduced peroxidation and nitration products indicative of free radical mediated tissue injury, specifically, malondialdehyde and nitrotyrosine. Reduced markers of DNA damage and superoxide dismutase were also observed in the bioactive RLX-treated group. RLX administration was also associated with reduced ileal cell apoptosis as measured by caspase 3 and terminal deoxynucleotidyltransferase-mediated UTP end labeling. These were in contrast to the findings in SAO/R rats treated with inactive RLX or vehicle. These data suggest that, similar to myocardial ischemia-reperfusion injury, RLX may reduce reperfusion-related tissue injury in the intestinal tract.

Relaxin and Liver Transplantation—Preservation and Reperfusion

Reperfusion injury is also major problem in organ transplant. Boehnert and colleagues studied liver transplantation using isolated perfused rat livers[112] The investigators administered rhRLX in the preservation solution or with both preservation and reperfusion. rhRLX 32 ng/ml was used for reperfusion and 64 ng/ml used for preservation. The period of ischemia was 3.5 hours with one group at 20°C and the other at 4°C. Cell damage was quantified by measuring malondialdehyde and myeloperoxidase in the perfusate as well as immunohistochemical staining of the liver. In both warm and cold ischemia, relaxin treatment reduced malondialdehyde and myeloperoxidase in the perfusate as well as in the tissue staining. These exciting preliminary data suggest that relaxin may have a role in preservation as well as the reperfusion phases of organ transplantation.

Relaxin and Pulmonary Hypertension

The vasodilatory and anti-fibrotic effects of relaxin are also proposed to ameliorate pulmonary hypertension. Tozzi and colleagues utilized a hypoxia model of pulmonary hypertension in rats.[113] Two doses of rhRLX (0.24 mg/kg and 0.05 mg/kg) were administered subcutaneously for 10 days to hypoxic (10% oxygen) rats. On day 11, RLX reduced right ventricular pressure in a dose-dependent manner in these rats anesthetized with pentobarbital. In the high dose relaxin group, collagen accumulation was reduced in the main pulmonary artery compared to untreated hypoxic controls. Right ventricular pressures were not significantly decreased with RLX administration to air-breathing rats In vitro studies of cultured rat pulmonary artery fibroblast cells revealed that relaxin reduced collagen and fibronectin in TGF-β stimulated cells. These investigators suggest that relaxin may suppress fibroproliferation in hypoxia-induced pulmonary hypertension and may have therapeutic benefit in some cases of pulmonary hypertension.

Relaxin and Vascular Inflammation

Neutrophil margination within blood vessels and endothelial expression of adhesion markers can result in endothelial and vascular dysfunction as well as extravasation of neutrophils with subsequent tissue injury. Nistri and colleagues have approached this issue from the perspective of myocardial ischemia and the potential role of relaxin in ameliorating this inflammation mediated tissue injury in the heart. To this end, rat coronary artery endothelial cells primed with lipopolysaccharide were cocultured in vitro with neutrophils.[88] Pretreatment with porcine relaxin reduced the adherent neutrophils by light microscopy compared to controls pretreated with inactivated RLX.

Surface adhesion molecules, P-selectin and VCAM-1, expression by Western blot, immunohistochemistry and PCR, were reduced with relaxin pretreatment. Administration of the nitric oxide synthase inhibitor, L-NMMA, significantly attenuated these salutary effects of relaxin. Thus, a nitric oxide mediated mechanism is implicated in the reduced vascular inflammation observed with RLX pretreatment. Masini and colleagues extended these findings to human neutrophils and demonstrated that RLX could inhibit activation of isolated human neutrophils.[87] Relaxin reduced surface expression of CD11b, reduced the generation of superoxide anion, reduced the rise in intracellular calcium and reduced the release of cytoplasmic granules and chemotactic migration. These relaxin-mediated effects were blunted with L-NMMA. Thus, relaxin-mediated effects may also play a role in pregnancy and counteract the excess maternal inflammatory response seen with certain pregnancy associated disorders, such as preeeclampsia and may have therapeutic implications in other disorders associated with excessive vascular inflammation. However, not all publications are consistent with this concept.[114]

Relaxin and Preeclampsia

Preeclampsia is a pregnancy-specific disorder that occurs during the second half of gestation. This syndrome is characterized by generalized vasoconstriction and endothelial dysfunction. During active disease, the serum immunoreactive relaxin concentrations are not significantly different from those measured in normotensive, gestational-aged matched controls.[115] However, whether bioactive serum concentrations may be decreased, or whether there may be reduced numbers of vascular relaxin receptors or a defect in postreceptor signaling during preeclampsia is unknown. Furthermore, in women destined to develop preeclampsia, the status of serum relaxin concentrations during the first trimester when peak levels are reached is also unknown. It is possible that abnormally low or high serum concentrations at that time may result in deficient or exaggerated maternal renal and cardiovascular adaptations that, in turn, predispose women to develop the disease.

By virtue of its renal vasodilatory attributes, relaxin administration could improve renal plasma flow and glomerular filtration rate in women with preeclampsia. Relaxin therapy would be expected to decrease systemic vascular resistance and increase global arterial compliance, thereby augmenting cardiac output and maternal organ perfusion. There is limited information indicating that relaxin may also be a uterine vasodilator.[52] Perhaps relaxin could enhance uteroplacental blood flow by dilating unremodelled spiral arteries containing vascular smooth muscle, thereby improving oxygenation of the intervillous space and attenuating placental expression of hypoxia-inducible transcription factors and their regulated genes such as the anti-angiogenic, soluble fms-like tyrosine kinase and soluble endoglin.[116-118] Because relaxin is primarily an arterial vasodilator, preload is not compromised and thus, cardiac output is reciprocally increased, such that undue hypotension should not be a limitation of relaxin therapy. Rather, blood pressure can be reduced using the standard anti-hypertensive agents as necessary for maternal health concerns.

One question is whether more relaxin is better in preeclampsia. As mentioned, immunoreactive serum concentrations are comparable in preeclamptic and normotensive controls, but for both, they are only ~50% of the peak levels observed in the first trimester of normal pregnancies. Ultimately, the answer to this question can only be determined by clinical investigation. Another question is whether the vasodilatory properties of relaxin that ultimately depend on endothelial NO production (at least in health) are preserved in the face of the "endothelial dysfunction" that accompanies preeclampsia. In partial answer to this question, relaxin is fully active in spontaneously hypertensive rats and angiotensin II-infused rats, two animal models of hypertension associated with endothelial dysfunction.[9]

Last, based on both circumstantial evidence and preliminary experiments from our laboratory, we hypothesize that vascular endothelial growth factor (VEGF) is an intermediary molecule in the relaxin vasodilatory pathway positioned between the relaxin receptor and gelatinase. On the one hand, circulating anti-angiogenic molecules such as sFlt-1 may impair the relaxin vasodilatory pathway, thereby contributing to the pathogenesis of preeclampsia. On the other, relaxin supplementation may restore endothelial cell health in the disease by enhancing local production of VEGF within the arterial wall, thereby partly or wholly neutralizing the deleterious effects of circulating anti-angiogenic factors. Finally, the anti-inflammatory properties of relaxin as reviewed above (Relaxin and vascular inflammation) may be beneficial in the setting of preeclampsia.

Part 3. Controversies, Unresolved Issues and Future Directions

Over the past decade, relaxin has emerged as a hormone with multiple vascular actions. However, there are a number of controversies or unresolved issues and some of the more glaring ones are summarized below. Hopefully other laboratories will weigh in on these controversies with additional data in the near future.

1. There are two, very different proposals to explain the vasodilatory action of relaxin. One advanced by Bani and colleagues implicates the induction of iNOS and the other by Conrad and coworkers implicates the activation of eNOS (supra vide).
2. The discrepant findings of Dschietzig and colleagues[35] and Kerchner et al[34] on whether Rlx upregulates the endothelial ET_B receptor remains unresolved.
3. Li and colleagues[119] recently claimed that chronic relaxin administration to rats did not reduce the myogenic reactivity of isolated mesenteric arteries and they attributed the apparent reductions observed by Novak and coworkers[27] to increased passive compliance. However, other explanations for the different results arising from the two laboratories are possible. First, Li et al investigated *myogenic tone* in which the main independent variable is the presence or absence of extracellular calcium (at any given pressure), while Novak and colleagues examined *myogenic reactivity* where the main independent variable is change in intraluminal pressure (in the presence of calcium). Therefore, the two studies may not be directly comparable. Second, Li and colleagues used small mesenteric arteries, while Novak and coworkers have focused on small renal arteries. There may be differences secondary to the anatomical origin of arteries. Third, the myogenic reactivity of small renal arteries isolated from midterm pregnant rats is reduced, yet passive compliance remains unchanged at this stage of pregnancy[10,11]. This reduction in myogenic reactivity is due to relaxin, because it was found to be reversed in small renal arteries isolated from midterm pregnant rats that received relaxin neutralizing antibodies in vivo.[5] Therefore, the reduction in myogenic reactivity due to relaxin is found to occur in the absence of any increase in passive compliance. Fourth, the reduction in myogenic reactivity of small renal arteries isolated from nonpregnant rats chronically administered rhRlx is reversed by short-term treatment with NO synthase inhibitors, ET_B receptor antagonists and MMP inhibitors.[27,36] It is highly unlikely that the increase in passive compliance which reflects a remodeling of the vascular wall will be reversed by these inhibitors in such a short period of time (30 min). A similar conclusion can be drawn for the recent investigations on arterial function in wild-type and relaxin knock-out mice in which short-term, 30 min incubations with L-arginine were used.[72]
4. There is not unanimity among the studies on whether serum relaxin is significantly increased in congestive heart failure.[73,101] Furthermore, despite using the same relaxin assay, serum relaxin concentrations were not uniform between the studies. The latter issue points to a general deficiency in the field, i.e., the availability of validated, reliable and sensitive, commercial assays for measuring human relaxin.

In view of the numerous and generally beneficial actions of relaxin in the vasculature learned mainly from animal investigations, there is tremendous potential for therapeutics. Indeed, the therapeutic potential of relaxin has been tested and validated in various animal models of human disease. Future investigations need to establish whether these beneficial vascular actions of relaxin translate to humans.

Acknowledgements

We gratefully acknowledge the invaluable contributions of our many collaborators over the years. The work from the authors' laboratory has been supported most recently by NIH RO1 HD30325,

DK63321 and HL67937. Dr. Jeyabalan is supported by the BIRCWH Faculty Development Award (NIH K12-HD043441-04) and Dr. Shroff by the Mcginnis Chair Endowment Funds

References

1. Dallenbach-Hellweg G, Dawson AB, Hisaw FL. The effect of relaxin on the endometrium of monkeys- histological and histochemical studies. Am J Anatomy 1966; 119:61-78.
2. Hisaw FL, Hisaw FL, Jr, Dawson AB. Effects of relaxin on the endothelium of endometrial blood vessels in monkeys (Macaca mulatta). Endocrinology 1967; 81:375-385.
3. Shroff SG. Pulsatile arterial load and cardiovasculat function: fact, fiction and wishful thinking. Therapeutic Research 1998; 19:59-66.
4. Poppas A, Shroff SG, Korcarz CE et al. Serial assessment of the cardiovascular system in normal pregnancy. Role of arterial compliance and pulsatile arterial load. Circulation 1997; 95:2407-2415.
5. Novak J, Danielson LA, Kerchner LJ et al. Relaxin is essential for renal vasodilation during pregnancy in conscious rats. J Clin Invest 2001; 107:1469-1475.
6. Conrad KP, Debrah DO, Novak J et al. Relaxin modulates systemic arterial resistance and compliance in conscious, nonpregnant rats. Endocrinology 2004; 145:3289-3296.
7. Shroff S, Berger D, Lang R et al. Physiologic relevance of T-Tube model parameters with emphasis on arterial compliances. Am J Physiol (Heart Circ Physiol 38) 1995; 269:H365-H374.
8. Debrah DO, Conrad KP, Danielson LA et al. Effects of relaxin on systemic arterial hemodynamics and mechanical properties in conscious rats: gender dependency and dose response. J Appl Physiol 2005; 98:1013-1020.
9. Debrah DO, Conrad KP, Jeyabalan A et al. Relaxin increases cardiac output and reduces systemic arterial load in hypertensive rats. Hypertension 2005; 46:745-750.
10. Debrah DO, Novak J, Matthews JE et al. Relaxin is essential for systemic vasodilation and increased global arterial compliance during early pregnancy in conscious rats. Endocrinol 2006; 147:5126-31.
11. Gandley RE, Conrad KP and McLaughlin MK. Endothelin and nitric oxide mediate reduced myogenic reactivity of small renal arteries from pregnant rats. Am J Physiol Regul Integr Comp Physiol 2001; 280:R1-7.
12. Gandley RE, Griggs KC, Conrad KP et al. Intrinsic tone and passive mechanics of isolated renal arteries from virgin and late pregnant rats. Am J Physiol 1997; 273:R22-R27.
13. McLaughlin MK, Keve TM. Pregnancy-induced changes in resistance blood vessels. Am J Obstet Gynecol 1986; 155:1296-1299.
14. Conrad KP, Lindheimer MD Renal and cardiovascular alterations. In: Lindheimer MD, Cunningham FG, Roberts JM eds. Appleton and Lange, Chesley's Hypertensive Disorders in Pregnancy: Second Edition. Chapter 8. Stamford, CT, 1999; 263-326.
15. Conrad KP Mechanisms of renal vasodilation and hyperfiltration during pregnancy. J Soc Gynecol Invest 2004; 11, 438-48.
16. Conrad KP, Novak J Emerging role of relaxin in renal and cardiovascular function. Am J Physiol Regul Integr Comp Physiol 2004; 287:4250-R261.
17. Davison JM, Noble MC. Serial changes in 24 hour creatinine clearance during normal menstrual cycles and the first trimester of pregnancy. Br J Obstet Gynaecol 1981; 88:10-7.
18. Baylis C The mechanism of the increase in glomerular filtration rate in the twelve-day pregnant rat. J Physiol 1982; 22:1982;136-45.
19. Roberts M, Lindheimer MD, Davison JM. Altered glomerular permselectivity to neutral dextrans and heteroporous membrane modeling in human pregnancy. Am J Physiol 1996; 270:F338-43.
20. Conrad KP. Renal hemodynamics during pregnancy in chronically catheterized, conscious rats. Kidney Int 1984; 26:24-9.
21. Sherwood OD Relaxin. In: The Physiology of Reproduction. Eds: Knobil E, Neill JD, Greenwald GS, Markert CL, Pfaff DW, Raven, NY 1994; 861-1009.
22. Danielson LA, Sherwood OD, Conrad KP. Relaxin is a potent renal vasodilator in conscious rats. J Clin Invest 1999; 103:525-33.
23. Danielson LA, Kerchner LJ, Conrad KP. Impact of gender and endothelin on renal vasodilation and hyperfiltration induced by relaxin in conscious rats. Am J Physiol Regul Integr Comp Physiol 2000; 279:R1298-304.
24. Conrad KP, Colpoys MC. Evidence against the hypothesis that prostaglandins are the vasodepressor agents of pregnancy. Serial studies in chronically instrumented conscious rats. J Clin Invest 1986; 77:236-45.
25. Danielson LA, Conrad KP. Acute blockade of nitric oxide synthase inhibits renal vasodilation and hyperfiltration during pregnancy in chronically instrumented, conscious rats. J Clin Invest 1995; 96:482-90.

26. Novak J, Reckelhoff J, Bumgarner L et al. Reduced sensitivity of the renal circulation to angiotensin II in pregnant rats. Hypertension 1997; 30:580-4.
27. Novak J, Ramirez RJ, Gandley RE et al. Myogenic reactivity is reduced in small renal arteries isolated from relaxin-treated rats. Am J Physiol Regul Integr Comp Physiol 2002; 283:R349-55.
28. Danielson LA, Conrad KP. Time course and dose response of relaxin-mediated renal vasodilation, hyperfiltration and changes in plasma osmolality in conscious rats. J App Physiol 2003; 95:1509-1514.
29. Smith MC, Danielson LA, Conrad KP et al. Influence of recombinant human relaxin on renal haemodynamics in humans. J Am Soc Nephrol 2006; 17:3192-7.
30. Erikson MS, Unemori EN. Relaxin clinical trials in systemic sclerosis In: Tregear GW, Ivell R, Bathgate RA et al. eds. Relaxin 2000. Kluwer Academic Publishers, MA 2000; 373-381.
31. Smith MC, Murdoch AP, Danielson LA et al. Relaxin has a role in establishing a renal response in pregnancy. Fertil Steril 2006; 86:253-255.
32. Novak J, Rajakumar A, Miles TM et al. Nitric oxide synthase isoforms in the rat kidney during pregnancy. J Soc Gynecol Invest 2004; 11:280-288.
33. Alexander BT, Miller MT, Kassab L et al. Differential Expression of Renal Nitric Oxide Synthase Isoforms During Pregnancy in Rats. Hypertension Journal of the American Heart Association 1999; 33:435-439.
34. Kerchner LJ, Novak J, Hanley-Yanez K et al. Evidence against the hypothesis that endothelial endothelin B receptor expression is regulated by relaxin and pregnancy. Endocrinol 2005; 146:2791-7.
35. Dschietzig T, Bartsch C, Richter C et al. Relaxin, a pregnancy hormone, is a functional endothelin-1 antagonist. Circ Res 2003; 92:32-40.
36. Jeyabalan A, Novak J, Danielson LA et al. Essential role for vascular gelatinase activity in relaxin-induced renal vasodilation, hyperfiltration and reduced myogenic reactivity of small arteries. Circ Res 2003; 93:1249-1257.
37. Novak J, Conrad KP. Small renal arteries isolated from ETB receptor deficient rats fail to exhibit the normal maternal adaptation to pregnancy. FASEB J 2004; 18(5) Part I, abstract #205.32.
38. Jeyabalan A, Kerchner LJ, Fisher MC et al. Matrix metalloproteinase-2 activity, protein, mRNA and tissue inhibitors in small arteries from pregnant and relaxin-treated nonpregnant rats. J Applied Physiol 2006; 100:1955-1963.
39. Jeyabalan A, Novak J, Doty KD et al. Vascular matrix metalloproteinase-9 mediates the inhibition of myogenic reactivity in small arteries isolated from rats after short term administration of relaxin. Endocrinology 2007; 189:-197.
40. Matthews JE, Rubin JP, Novak J et al. Vascular Endothelial Growth Factor (VEGF) Is a New Player In Slow Relaxin (Rlx) Vasodilatory Pathway. Reproductive SCI 2007; 14(1 suppl):114A.
41. Fisher C, MacLean M, Morecroft I et al. Is the pregnancy hormone relaxin also a vasodilator peptide secreted by the heart? Circulation 2002; 106:292-295.
42. Matthews JE, Rubin JP, Novak J et al. Relaxin (Rlx) Induces Fast Relaxation In some Rat and Human Arteries Mediated By P13 Kinase And Nitric Oxide. Reproductive SCI 2007; 14(1 suppl):114A.
43. Bani D, Failli P, Bello MG et al. Relaxin activates the L-arginine-nitric oxide pathway in vascular smooth muscle cells in culture. Hypertension 1998; 31:1240-1247.
44. Failli P, Nistri S, Quattrone S et al. Relaxin up-regulates inducible nitric oxide synthase expression and nitric oxide generation in rat coronary endothelial cells. FASEB Journal 2002; 16:252-254.
45. Quattrone S, Chiappini L, Scapagnini G et al. Relaxin potentiates the expression of inducible nitric oxide synthase by endothelial cells from human umbilical vien in in vitro culture. Molecular Hum Reprod 2004; Vol.10, No.5:325-330.
46. Sladek SM, Magness RR, Conrad KP. Nitric oxide and pregnancy. Am J Physiol 1997; 272: R441-463.
47. Conrad KP, Gandley RE, Ogawa T et al. Endothelin mediates renal vasodilation and hyperfiltration during pregnancy in chronically instrumented conscious rats. Am J Physiol Renal Physiol 1999; 276: F767-776.
48. Bani ST, Bigazzi M, Bani D et al. Relaxin-induced increased coronary flow through stimulation of nitric oxide production. Br J Pharmacol 1995; 116:1589-1594.
49. Vasilenko P, Mead JP, Weidmann JE et al. Uterine growth-promoting effects of relaxin: a morphometric and histological analysis. Biol Reprod 1986; 35:987-995.
50. Kohsaka T, Min G, Lukas G et al. Identification of specific relaxin-binding cells in the human female. Biol Reprod 1998; 59:991-999.
51. Min G, Sherwood OD. Identification of specific relaxin-binding sites in the cervix, mammary glands, nipples, small intestine and skin of pregnant pigs. Biol Reprod 1996; 55:1243-1252.
52. Novak J. Relaxin increases uterine blood flow in conscious nonpregnant rats and decreases myogenic reactivity in isolated uterine arteries. FASEB J 2002; 16:A824.

53. Schramm W, Einer-Jensen N, Brown MB et al. Effect of four primary prostaglandins and relaxin on blood flow in the ovine endometrium and myometrium. Biol Reprod 1984; 30:523-531.
54. Roche PJ, Crawford RJ, Tregear GW. A single-copy relaxin-like gene sequence is present in sheep. Mol Cell Endocrinol 1993; 91:21-28.
55. Unemori EN, Erikson ME, Rocco SE et al. Relaxin stimulates expression of vascular endothelial growth factor in normal human endometrial cells in vitro and is associated with menometrorrhagia in women. Hum Reprod 1999; 14:800-806.
56. Jauniaux E, Johnson MR, Jurkovic D et al. The role of relaxin in the development of the uteroplacental circulation in early pregnancy. Obstet Gynecol 1994; 84:338-342.
57. Petersen LK, Svane D, Uldbjerg N et al. Effects of human relaxin on isolated rat and human myometrium and uteroplacental arteries. Obstet Gynecol 1991; 78:757-762.
58. Dombrowski MP, Savoy-Moore RT, Swartz K et al. Effect of porcine relaxin on the human umbilical artery. J Reprod Med 1986; 31:467-472.
59. Bani G, Bani TS, Bigazzi M et al. Effects of relaxin on the micorvasculature of mouse mammary gland. Histol Histopath 1988; 3:337-343.
60. Bigazzi M, Bani G, Bani ST et al. Relaxin: a mammotropic hormone promoting growth and differentiation of the pigeon crop sac mucosa. Acta Endocrinologica 1988; 117:181-188.
61. Bani D, Nistri S, Quattrone S et al. The vasorelaxant hormone relaxin induces changes in liver sinusoid microcirculation: a morphologic study in the rat. J Endocrinol 2001; 171:541-549.
62. Massicotte G, Parent A, St-Louis J. Blunted responses to vasoconstrictors in mesenteric vasculature but not in portal vein of spontaneously hypertensive rats treated with relaxin. Proc Soc Exp Biol Med 1989; 190:254-259.
63. Bigazzi M, Del Mese A, Petrucci F et al. The local administration of relaxin induces changes in the microcirculation of the rat mesocaecum. Acta Endocrinologica 1986; 112:296-299.
64. Goldsmith LT, Weiss G, Palejwala S et al. Relaxin regulation of endometrial structure and function in the rhesus monkey. PNAS 2004; 101(13).
65. Palejwala S, Tseng L, Wojtczuk A et al. Relaxin Gene and Protein Expression and its Regulation of Procollagenase and Vascular Endothelial Growth Factor in Human Endometrial Cells. Biol Reprod 2002; 66:1743-1748.
66. Shirota K, Tateishi K, Emotok et al. Relaxin-induced angiogenesis in ovary contributes to follicle development. Ann NY Acad Sci. 2005; 1041:144-6.
67. Silvertown JD, Ng J, Sato T et al. H2 relaxin overexpression increases in vivo prostate xenograft tumor growth and angiogenesis. Int J Cancer 2006; 118:62-73.
68. Unemori EN, Lewis M, Constant J et al. Relaxin induces vascular endothelial growth factor expression and angiogenesis selectively at wound sites. Wound Repair Regen 2000; 8:361-370.
69. Norrby K, Bani D, Bigazzi M et al. Relaxin, a potent microcirculatory effector, is not angiogenic. Int J Microcirc 1996; 16:227-231.
70. Huang X, Arnold G, Lewis M et al. Effect of relaxin on normal and impaired wound healing in rodents.In: Tregear GW, Ivell, R., Bathgate, RA, Wade, JD Dordrecht, eds. Relaxin 2000: Proceedings of the third international conference on relaxin and related peptides, The Netherlands: Kluwer Academic Publishers 2001; 393-397.
71. Lewis M, Deshpande U, Guzman L et al. Systemic relaxin administration stimulates angiogenic cytokine expression and vessel formation in a rat myocardial infarct model. In: Tregear GW, Ivell, R, Bathgate, RA, Wade, JD Dordrecht, eds. Relaxin 2000: Proceedings of the third international conference on relaxin and related peptides, The Netherlands: Kluwer Academic Publishers, 2001; 159-167.
72. Novak J, Parry LJ, Matthews JE et al. Evidence for local relaxin ligand-receptor expression and function in arteries. FASEB J 20, 2006; 2352-2362.
73. Dschietzig T, Richter C, Bartsch C et al. The pregnancy hormone relaxin is a player in human heart failure. FASEB Journal 2001; 15:2187-2195.
74. Palejwala S, Stein DE, Weiss G et al. Relaxin positively regulates matrix metalloproteinase expression in human lower uterine segment fibroblasts using a tyrosine kinase signaling pathway. Endocrinology 2001; 142:3405-3413.
75. Unemori EN, Amento EP. Relaxin modulates synthesis and secretion of procollagenase and collagen by human dermal fibroblasts. J Biol Chem 1990; 265:10681-10685.
76. Unemori EN, Pickford LB, Salles AL et al. Relaxin induces an extracellular matrix-degrading phenotype in human lung fibroblasts in vitro and inhibits lung fibrosis in a murine model in vivo. J Clin Invest 1996; 98:2739-2745.
77. Samuel CS, Zhao C, Bond CP et al. Relaxin-1-deficient mice develop an age-related progression of renal fibrosis. Kidney Int 2004; 65:2054-2064.
78. Garber SL, Mirochnik Y, Brecklin CS et al. Relaxin decreases renal interstitial fibrosis and slows progression of renal disease. Kidney Int 2001; 59:876-882.

79. Garber SL, Mirochnik Y, Brecklin C et al. Effect of relaxin in two models of renal mass reduction. Am J Nephrol 2003; 23:8-12.
80. Huang X, Cheng Z, Sunga J et al. Systemic administration of recombinant human relaxin (RHRLX) ameliorates the acute cyclosporine nephrotoxicity in rats. J Heart Lung Transplant 2001; 20:253.
81. McDonald GA, Sarkar P, Rennke H et al. Relaxin increases ubiquitin-dependent degradation of fibronectin in vitro and ameliorates renal fibrosis in vivo. Am J Physiol Renal Physiol 2003; 285(1):F59-67.
82. Samuel CS, Hewitson TD. Relaxin in cardiovascular and renal disease. Kidney Int 2006; 69(9):1498-502.
83. Danielson LA, Welford A, Harris A. Relaxin Improves Renal Function and Histology in Aging Munich Wistar Rats. J Am Soc Nephrol 2006; 17(5):1325-33.
84. Taylor MJ, Clark, CL. Evidence for a novel source of relaxin: atrial cardiocytes. J Endocrinol 1994; 143:R5-R8.
85. Masini E, Bani D, Bello MG et al. Relaxin counteracts myocardial damage induced by ischemia-reperfusion in isolated guinea pig hearts: evidence for an involvement of nitric oxide. Endocrinology 1997; 138:4713-4720.
86. Bani D, Masini E, Bello MG et al. Relaxin protects against myocardial injury caused by ischemia and reperfusion in rat heart. Am J Pathol 1998; 152:1367-1376.
87. Masini E, Nistri S, Vannacci A et al. Relaxin inhibits the activation of human neutrophils: involvement of the nitric oxide pathway. Endocrinology 2004; 145(3):1106-12.
88. Nistri S, Chiappini L, Sassoli C et al. Relaxin inhibits lipopolysaccharide-induced adhesion of neutrophils to coronary endothelial cells by a nitric oxide-mediated mechanism. FASEB Journal 2003; 17(14):2109-11.
89. Bani D, Ballati L, Masini E et al. Relaxin counteracts asthma-like reaction induced by inhaled antigen in sensitized guinea pigs. Endocrinology 1997; 138:1909-1915.
90. Masini E, Bani D, Bigazzi M et al. Effects of relaxin on mast cells: In vitro and in vivo studies in rats and guinea pigs. J Clin Invest 1994; 94:1974-1980.
91. Bani D, Bigazzi M, Masini E et al. Relaxin depresses platelet aggregation: in vitro studies on isolated human and rabbit platelets. Lab Invest 1995; 73:709-716.
92. Zhang J, Qi YF, Geng B et al. Effect of relaxin on myocardial ischemia injury induced by isoproterenol. Peptides 2005; 26(9):1632-9.
93. Perna AM, Masini E, Nistri S. Human recombinant relaxin reduces heart injury and improves ventricular performance in a swine model of acute myocardial infarction. Ann N Y Acad Sci 2005; 1041:431-3.
94. Bani D, Nistri S, Sacchi TB et al. Basic progress and future therapeutic perspectives of relaxin in ischemic heart disease. Ann N Y Acad Sci 2005; 1041:423-30.
95. Du X-J, Samuel CS, Gao X-M et al. Increased myocardial collagen and ventricular diastolic dysfunction in relaxin deficient mice: a gender-specific phenotype. Cardiovasc Res 2003; 57:395-404.
96. Samuel CS, Unemori EN, Mookerjee I et al. Relaxin modulates cardiac fibroblast proliferation, differentiation and collagen production and reverses cardiac fibrosis in vivo. Endocrinology 2004; 145(9):4125-33.
97. Kompa AR, Samuel CS, Summers RJ. Inotropic responses to human gene 2 (B29) relaxin in a rat model of myocardial infarction (MI): effect of pertussis toxin. Br J Pharmacol 2002; 137:710-718.
98. Osheroff PL, King KL. Binding and cross-linking of 32P-labeled human relaxin to human uterine cells and primary rat atrial cardiomyocytes. Endocrinology 1995; 136:4377-4381.
99. Lekgabe ED, Kiriazis H, Zhao C et al. Relaxin reverses cardiac and renal fibrosis in spontaneously hypertensive rats. Hypertension 2005; 46(2):412-8.
100. Dschietzig T, Bartsch C, Kinkel T et al. Myocardial relaxin counteracts hypertrophy in hypertensive rats. Ann N Y Acad Sci 2005; 1041:441-3.
101. Krüger S, Graf J, Merx MW et al. Relaxin kinetics during dynamic exercise in patients with chronic heart failure. Eur J Intern Med 2004; 15:4-56.
102. Hassink RJ, Brutel de la Riviere A, Mummery CL et al. Transplantation of cells for cardiac repair. J Am Coll Cardiol 2003; 41(5):711-7.
103. Dowell JD, Rubart M, Pasumarthi KB et al. Myocyte and myogenic stem cell transplantation in the heart. Cardiovasc Res 2003; 58(2):336-50.
104. Murry CE, Wiseman RW, Schwartz SM et al. Myoblast transplantation for repair of myocardial necrosis. J Clin Invest 1996; 98(11):2512-23.
105. Taylor DA, Atkins BZ, Hungspreugs P et al. Regenerating functional myocardium: improved performance after skeletal myoblast transplantation. [erratum appears in Nat Med 1998; 4(10):1200]. Nat Med 1998; 4(8):929-33.
106. Formigli L, Francini F, Chiappini L et al. Relaxin favors the morphofunctional integration between skeletal myoblasts and adult cardiomyocytes in coculture. Ann N Y Acad Sci 2005; 1041:444-5.

107. Masini E, Zagli G, Ndisang JF et al. Protective effect of relaxin in cardiac anaphylaxis: involvement of the nitric oxide pathway. Brit J Pharmacol 2002; 137:337-344.
108. Wilson BC, Milne P, Saleh TM. Relaxin pretreatment decreases infarct size in male rats after middle cerebral artery occlusion. Ann N Y Acad Sci 2005; 1041:223-8.
109. Nistri S, Bani D. Relaxin in vascular physiology and pathophysiology: possible implications in ischemic brain disease. Curr Neurovasc Res 2005; 2(3):225-3.
110. Wilson BC, Connell B, Saleh TM. Relaxin-induced reduction of infarct size in male rats receiving MCAO is dependent on nitric oxide synthesis and not estrogenic mechanisms. Neurosci Lett 2006; 393(2-3):160-4.
111. Masini E, Cuzzocrea S, Mazzon E et al. Protective effects of relaxin in ischemia/reperfusion-induced intestinal injury due to splanchnic artery occlusion. Br J Pharmacol 2006; 148:1124-32.
112. Boehnert MU, Hilbig H, Armbruster FP. Relaxin as an additional protective substance in preserving and reperfusion solution for liver transplantation, shown in a model of isolated perfused rat liver. Ann N Y Acad Sci 2005; 1041:434-40.
113. Tozzi CA, Poiani GJ, McHugh NA et al. Recombinant human relaxin reduces hypoxic pulmonary hypertension in the rat. Pulm Pharmacol Ther 2005; 18(5):346-53.
114. Figueiredo KA, Mui AL, Nelson CC et al. Relaxin stimulates leukocyte adhesion and migration through a relaxin receptor LGR7-dependent mechanism. J Biol Chem 2006; 281(6):3030-9.
115. Szlachter BN, Quagliarello J, Jewelewicz R et al. Relaxin in normal and pathogenic pregnancies. Obstet Gynecol 1982; 59(2):167-70.
116. Rajakumar A, Brandon HM, Daftary A et al. Evidence for the functional activity of hypoxia-inducible transcription factors overexpressed in preeclamptic placentae. Placenta 2004; 25(10):763-9.
117. Maynard SE, Min JY, Merchan J et al. Excess placental soluble fms-like tyrosine kinase 1 (sFlt1) may contribute to endothelial dysfunction, hypertension and proteinuria in preeclampsia. J Clin Invest. 2003; 111(5):600-2.
118. Venkatesha S, Toporsian M, Lam C et al. Soluble endoglin contributes to the pathogenesis of preeclampsia. Nat Med. 2006; 12(6); 642-9.
119. Li Y, Brookes Z, Kaufman S. Acute and chronic effects of relaxin on vasoactivity, myogenic reactivity and compliance of the rat mesenteric arterial and venous vasculature. Regul Pept 2005; 132 41-46.

Chapter 7

The Effects of Relaxin on Extracellular Matrix Remodeling in Health and Fibrotic Disease

Chrishan S. Samuel,* Edna D. Lekgabe and Ishanee Mookerjee

Abstract

Since its discovery as a reproductive hormone 80 years ago, relaxin has been implicated in a number of pregnancy-related functions involving extracellular matrix (ECM) turnover and collagen degradation. It is now becoming evident that relaxin's ability to reduce matrix synthesis and increase ECM degradation has important implications in several nonreproductive organs, including the heart, lung, kidney, liver and skin. The identification of relaxin and RXFP1 (Relaxin family peptide receptor-1) mRNA and/or binding sites in cells or vessels of these nonreproductive tissues, has confirmed them as targets for relaxin binding and activity. Recent studies on *Rln1* and *Rxfp1* gene-knockout mice have established relaxin as an important naturally occurring and protective moderator of collagen turnover, leading to improved organ structure and function. Furthermore, through its ability to regulate the ECM and in particular, collagen at multiple levels, relaxin has emerged as a potent anti-fibrotic therapy, with rapid-occurring efficacy. It not only prevents fibrogenesis, but also reduces established scarring (fibrosis), which is a leading cause of organ failure and affects several tissues regardless of etiology. This chapter will summarize these coherent findings as a means of highlighting the significance and therapeutic potential of relaxin.

Introduction

Relaxin has long been regarded as a hormone of pregnancy, since its discovery almost eighty years ago (see 1, 2 for review). In many mammalian species, relaxin is predominantly produced by the ovary and/or placenta during pregnancy and promotes a number of actions within the female reproductive tract to facilitate parturition. In males, relaxin is expressed in the prostate and testis, secreted into the seminal fluid and promotes several other actions. Humans (and higher primates) have three relaxin genes, termed *RLN1*,[3] *RLN2*[4] and *RLN3*,[5] of which the product of the *RLN2* gene, H2 relaxin, is the major stored and circulating form. Other mammals have two relaxin genes, designated *RLN1* (equivalent to human *RLN2*) and *RLN3* (equivalent to human *RLN3*). Since the 1950's, research into relaxin has been able to progressively demonstrate that the peptide hormone is a potent regulator of the extracellular matrix (ECM) and in particular, collagen turnover, an effect which has subsequently been identified in many reproductive and nonreproductive tissues. The identification of relaxin binding sites in the heart and brain provided further evidence that relaxin influences organs outside the reproductive tract. Furthermore, the recent discovery of the relaxin receptor as a leucine rich repeat-containing G-protein coupled receptor-7 (LGR7)[6]

*Corresponding Author: Chrishan S. Samuel—Howard Florey Institute, University of Melbourne, Parkville, Victoria, 3010, Australia. Email: chrishan.samuel@florey.edu.au

Relaxin and Related Peptides, edited by Alexander I. Agoulnik. ©2007 Landes Bioscience and Springer Science+Business Media.

(which has now been termed RXFP1), in addition to other relaxin family peptide receptors[7,8] that are activated by relaxin and/or relaxin-3 in several tissues, have confirmed that these peptides are able to bind to multiple target organs within the body and exert a number of diverse roles. In this chapter, we will specifically discuss the matrix remodeling and anti-fibrotic effects of these relaxin peptides, which have significant therapeutic and clinical implications.

The Extracellular Matrix (ECM)

The ECM of connective tissues can be defined as an intricate network of macromolecules, composed of a variety of insoluble fibres, microfibrils, proteins and polysaccharides that are secreted locally and assembled into an organized meshwork in close association with the surface of the cells that produce them.[9,10] These specialized macromolecules, which form unique compositions in each connective tissue of the body, play an integral role in regulating cell function, differentiation and tissue-specific gene expression[11] and are essential for maintaining the structure and function of tissue during development, tissue remodeling and repair processes. The primary role of the ECM is to provide tissues with its specific mechanical and physicochemical properties, in addition to a structural framework for cell attachment and migration.[12] Additionally, the ECM can also act as a regulator of cell growth and survival, activate intracellular signaling pathways and modulate the activity of a number of growth factors and proteins.[13] However, while the macromolecules within the ECM contain the necessary information required to promote their assembly and interaction, it is the ECM-producing cells that play a pivotal role by synthesizing and regulating this assembly, through specific interactions that occur between the cell receptors and matrix proteins.[14]

The predominant proteins of the ECM within connective tissues and in fact the body are members of the collagen glycoprotein family, of which over twenty different structural types exist.[15] Of the different types of collagen, type I collagen is the major collagenous form in most organs, where it forms characteristic fibre bundles to determine the maximal volume and integrity of these tissues. All collagens contain globular domains and share the common structural motif of triple helical segments. This triple helical structure is composed of three α– (polypeptide) chains which range from 10-150kDa per chain. Each chain consists of a repeating triplet mino acid sequence $(Gly\text{-}X\text{-}Y)_n$, where X and Y can be any amino acid, but are often proline and hydroxyproline, respectively.[9,10] Collagens are synthesized as soluble procollagens that are converted into their mature form by specific cleavage of terminal propeptides,[16] before they undergo further posttranslational modifications, extracellular processing, fibril formation and cross-linking to provide tissue with their tensile properties, while providing a scaffolding for cell attachment and migration.[15]

Fibroblasts

Fibroblasts are pluripotent cells of mesodermal origin that synthesize and secrete components of the ECM, while their ability to produce ECM components such as collagen is enhanced in the presence of pro-fibrotic stimuli, such as transforming growth factor beta (TGF-β). Upon repeated pathological stimuli, fibroblasts proliferate and differentiate (or activate) into myofibroblasts (which are cells that possess properties of both fibroblasts and smooth muscle cells), which in turn, further interact with the ECM and produce prodigious amounts of collagen (and other matrix proteins),[17,18] which leads to further tissue injury and scarring (fibrosis). TGF-β not only induces myofibroblast differentiation and ECM/collagen growth, but also protects these cells from nitric oxide mediated apoptosis.[18] Elevated numbers of myofibroblasts are present in fibrotic lesions during the active period of fibrosis, which is characterized by their expression of the protein, α-smooth muscle actin (α-SMA) expression. The rate of fibrosis is dependent upon the balance between myofibroblast accumulation and apoptosis.[18]

Fibrosis

Fibrosis is the final common pathway in numerous organ pathologies (Fig. 1) and can be characterized by the accumulation of excess fibrous connective tissue, mainly fibrillar collagen, in the extracellular matrix (ECM).[17] It causes hardening and stiffening of affected organs and

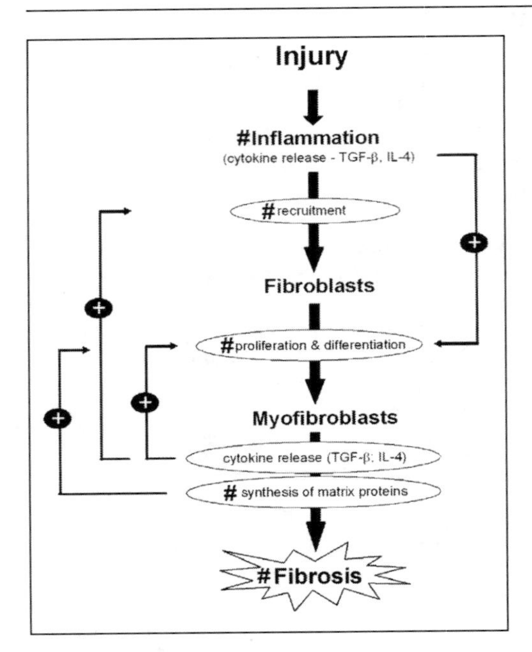

Figure 1. A generalized sequence of events leading from tissue injury to fibrosis. # indicates the areas at which relaxin has previously been demonstrated to act, to induce its anti-inflammatory and anti-fibrotic effects.

can result in the disruption of normal architecture and loss of organ function which can lead to several adverse outcomes that bear pathological and clinical significance. Fibrosis harshly affects the performance of several organs and is often life-threatening, while currently available therapies are ineffective in ameliorating its development and progression. There are two common types of fibrosis: i) *reparative fibrosis* where collagen deposition occurs as a consequence of tissue injury or cell death resulting in scar formation; and ii) *reactive fibrosis* where collagen deposition occurs due to direct stimulation of fibroblasts without cell injury and death. The underlying pathology of fibrogenesis is comparable in all organs hence insights into its mechanisms have important implications for our understanding of the pathogenesis of fibrosis in all organs.

Resident fibroblasts, located in the interstitial space, play a pivotal role in fibrogenesis and are the primary source of de novo synthesis of collagen. In healthy organs, fibroblasts are 'quiescent' hence fibrillar collagens are expressed at low levels and are maintained in this state by a balance between collagen synthesis and collagen degradation, which is controlled by ECM-degrading enzymes, the matrix metalloproteinases (MMPs). Following injury or disease, fibrogenesis occurs as a result of fibroblast recruitment and activation (into myofibroblasts). Furthermore, an imbalance between collagen degrading enzymes, the MMPs and their inhibitors, the tissue inhibitors of MMPs (TIMPs), can also contribute to the excessive collagen deposition and resultant fibrosis, with a fine balance between these mechanisms ultimately determining the extent of scarring.[19]

The Matrix Remodeling Effects of Exogenous Relaxin

Pregnancy-Related Effects of Exogenous Relaxin

Following the discovery of relaxin, several reports emerged on relaxin's pivotal role in facilitating the extensive ECM remodeling that occurs in target tissues such as the pubic symphysis, cervix, vagina, mammary glands and nipples during mammalian pregnancy (reviewed in refs. 1, 2). These physiological changes enable the safe delivery of young and are believed to be partly mediated by relaxin's ability to modify the density and quantity of collagen fibres within the ECM.

It was in fact relaxin's ability to transform the pubic symphysis from a relatively stiff inflexible fibrocartilage to a flexible and elastic interpubic ligament in guinea pigs that led to the discovery of the hormone.[20] Subsequent to this finding, relaxin was found to increase the amount of soluble collagen and decrease the insoluble collagen in relation to the total collagen in the mouse pubic symphysis.[21] Similarly, treatment of ovariectomized rats with relaxin was demonstrated to induce a reduction in pubic symphysial collagen content which was significantly enhanced with estrogen and totally antagonized with progesterone treatment.[22] Furthermore, ligament growth could be induced in an estrogen-primed guinea pig following the administration of relaxin,[23] consistent with relaxin's ability to stimulate ECM/collagen remodeling within the mammalian pelvic girdle during pregnancy. In separate studies, relaxin was also shown to play important roles in promoting successful parturition,[24] the control of myometrial activity[25] and cervical ripening,[26] when administered to ovariectomized pregnant rats. The increased extensibility of the rat cervix was attributed to relaxin's ability to increase collagen solubility and hyaluronic acid concentration, while decreasing total collagen concentration.[27]

Antifibrotic Effects of Exogenous Relaxin

Recently the ECM remodeling effects of relaxin have been extended beyond the reproductive field. Recombinant H2 relaxin, which is biologically active in rodents,[28] has been clearly shown to have the ability to inhibit collagen accumulation (fibrosis) associated with tissue damage via multiple mechanisms in a variety of in vitro and in vivo models of chemically or surgically induced fibrosis (reviewed in refs. 1,2,19,29; see Table 1). These combined studies have identified relaxin as an emerging novel therapeutic approach to target fibrosis in several nonreproductive organs.

Recombinant H2 relaxin has been reported to inhibit collagen synthesis and promote collagen degradation in several in vitro experiments performed. Several investigators have demonstrated that recombinant H2 relaxin reduces the over-expression of collagen synthesis and deposition by increasing expression and activation of MMPs (collagen degrading enzymes) and decreasing the expression of TIMPs in transforming growth factor-β-stimulated human dermal fibroblasts,[30] human pulmonary fibroblasts,[31] human uterine fibroblasts,[32] human renal fibroblasts,[33] rat hepatic stellate cells,[34,35] obstructed rat renal cortical fibroblasts[36] and rat atrial and ventricular fibroblasts (Fig. 2).[37] Relaxin was also shown to act in synergy with interferon-gamma (IFN-γ) to reduced collagen over-expression associated with human scleroderma fibroblasts.[38] Furthermore, relaxin was shown to inhibit TGF-β or angiotensin-II or insulin growth factor-I-induced fibroblast proliferation,[37] differentiation[36,37] and collagen-I lattice contraction[36] as a means of down-regulating the stimulated collagen expression. In all these studies though, relaxin did not affect basal/unstimulated collagen expression, suggesting that relaxin does not directly inhibit the expression of collagen.

Human recombinant H2 relaxin has also been demonstrated to successfully reverse collagen accumulation in several organs, such as the skin, lung, liver, kidney and heart of every in vivo model of induced fibrosis studied to date (see Table 1). Relaxin significantly inhibited fibrosis in a rat model of dermal scarring, induced by polyvinyl alcohol sponge implants and in a mouse model of dermal fibrosis, induced by capsule formation around implanted osmotic mini-pumps.[39] When administered over a 2-week period to a bleomycin-induced model of pulmonary fibrosis in mice, recombinant H2 relaxin restored lung injury-induced collagen accumulation and restored lung function to that observed in normal mice.[31] In a separate study, similar findings were also observed when H2 relaxin was administered over a 4-week period to a carbon tetrachloride-induced model of hepatic fibrosis.[34] Relaxin has also been shown to reduce renal scarring and TGF-β expression, while improving renal function (glomerular filtration rate), when administered to a bromoethylamine-induced model of chronic capillary necrosis in the rat,[40] in addition to two rat models of decreased renal mass.[41] Relaxin also increased ubiquitin-dependent degradation of fibronectin as a means of ameliorating renal fibrosis.[42] More recently, recombinant (H2) relaxin was shown to reverse established cardiac fibrosis in mice with cardiac-restricted over-expression of β2-adrenergic receptors (which undergo heart failure and premature death)[37] and in spontaneously hypertensive rats[43] over a two-week treatment period. While most of these studies have focused on the anti-fibrotic effects of H2 relaxin, limited studies have also shown that relaxin-3 can also attenuate isoproterenol-induced myocardial fibrosis (in the rat) and subsequently improve left

Table 1. Rodent models of fibrosis which have been used to evaluate the anti-fibrotic effects of relaxin

Organ Affected	Species	Model of Fibrosis Used	Type of Relaxin Used	Treatment Period	% Fibrosis Inhibited	Reference
Skin	Rat	•Fibrotic infiltration of polyvinyl alcohol sponge implants	Recombinant H2	2-weeks	25-29	39
	Mouse	•Fibrotic capsule formation around implanted osmotic mini-pumps	Recombinant H2	2-weeks	25	39
		•Relaxin knockout: age-related progressive fibrosis	Recombinant H2	2-weeks	95-100 (6-mo); no effect (12-mo)	60
Lung	Mouse	•Bleomycin-induced model of pulmonary fibrosis	Recombinant H2	2-weeks	65-100	31
		•Relaxin knockout: age-related progressive fibrosis	Recombinant H2	2-weeks	95-100 (9-mo); 70 (12-mo)	58
Liver	Rat	•Carbon-tetrachloride-induced model of hepatic fibrosis	Recombinant H2	4-weeks	60-65	34
Kidney	Rat	•Bromoethylamine-induced model of chronic capillary necrosis	Recombinant H2	4-weeks	75	40
		•Renal mass reduction by surgical excision or infarction-induced hypertension and glomerulosclerosis	Recombinant H2	4-weeks	36-45	41
		•Anti-glomerular basement membrane nephritis/fibrosis	Recombinant H2	4-weeks	43-80	42
		•Spontaneous hypertension-induced fibrosis	Recombinant H2	2-weeks	80-90 (10-mo)	43

continued on next page

Table 1. Continued

Organ Affected	Species	Model of Fibrosis Used	Type of Relaxin Used	Treatment Period	% Fibrosis Inhibited	Reference
	Mouse	•Relaxin knockout: age-related progressive fibrosis	Recombinant H2	2-weeks	52-62 (12-mo)	59
Heart	Rat	•Spontaneous hypertension (and age)-induced fibrosis	Recombinant H2	2-weeks	90-100 (10-mo)	43
		•Isoproterenol-induced ischemic injury-associated fibrosis	Synthetic relaxin-3	10-days	33	44
	Mouse	•Overexpression of β2-adrenergic receptor-induced fibrosis	Recombinant H2	2-weeks	55-60	37
		•Relaxin knockout: age-related progressive fibrosis	Recombinant H2	2-weeks	40 (12-mo)	37

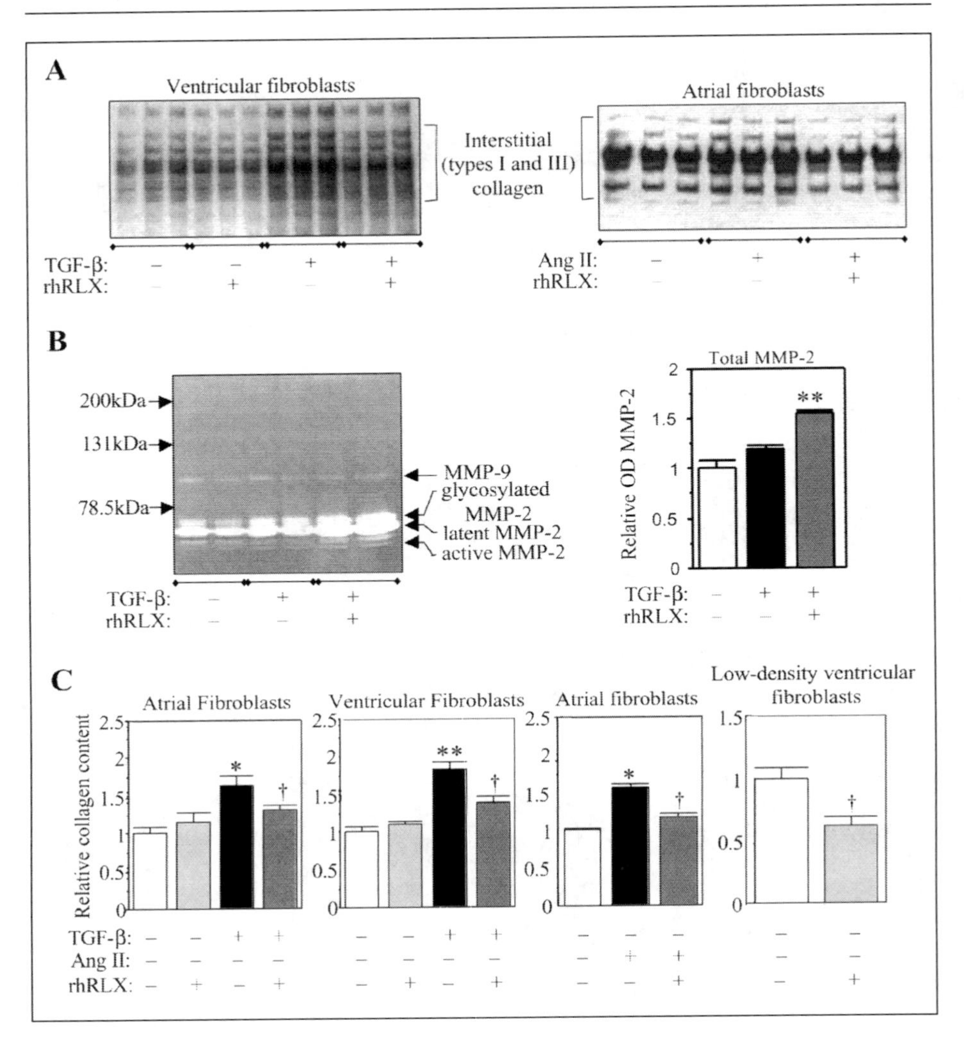

Figure 2. Modulation of collagen synthesis, degradation and deposition by recombinant H2 relaxin. Biosynthetically labelled interstitial collagens (A) from untreated cardiac fibroblasts ($2 \times 10^5/cm^2$) and cells treated with either recombinant H2 relaxin (100 ng/ml) alone, TGF-β (1 ng/ml) alone or TGF-β (1 ng/ml) and recombinant H2 relaxin (100 ng/ml), or with Ang II (5×10^{-7} M) alone or Ang II (5×10^{-7} M) and recombinant H2 relaxin (100 ng/ml), were measured from the media samples after 72 hours of culture. Shown are representative figures of triplicate samples from three separate experiments. MMP-2 and -9 expression and activity were determined by gelatin zymography (B) of media from untreated cultures and cells treated with either TGF-β (2 ng/ml) or TGF-β (2 ng/ml) and recombinant H2 relaxin (100 ng/ml) over 72 hours. Shown is a representative zymograph of duplicate samples from each group, from four sets of samples/group. Also shown are the mean ± SE 'relative OD MMP-2' of the total MMP-2 (derived from the latent and active forms of MMP-2), as determined by densitometry scanning. *Figure 2 legend coninued on next page.*

Figure 2 legend continued. Collagen content of cell layers (C) from untreated fibroblasts and cells treated with recombinant H2 relaxin (100 ng/ml) alone, TGF-β (2 ng/ml) alone or TGF-β (2 ng/ml) and recombinant H2 relaxin (100 ng/ml) or from untreated atrial fibroblasts and cells treated with Ang II (10^{-7} M) alone or Ang II (10^{-7} M) and recombinant H2 relaxin (100 ng/ml), after 72 hours of culture were also measured. Results are presented as the mean ± SE 'relative collagen content' from 3-4 separate experiments. $*P < 0.05$ and $**P < 0.01$ compared with values from untreated cells. $†P < 0.05$ compared with values from TGF-β or Ang II treated cells. Additionally, recombinant H2 relaxin (100 ng/ml)-treatment of low-density cells (5/mm^2) over 7 days caused an inhibition of collagen deposition. Results are presented as the mean ± SE 'relative collagen content' from 3 separate experiments (6 assays per group from each experiment). $†P < 0.05$ compared with values from untreated cells. Reproduced from reference 37 with permission; Copyright 2004, The Endocrine Society.

ventricular end-diastolic pressure following injury, while lowering plasma endothelin levels.[44] Additionally, H3 relaxin[45] and mouse relaxin-3[46] have been shown to stimulate MMP expression upon administration to rat cardiac fibroblasts and the mouse lung, respectively, consistent with a matrix-remodeling effect of this peptide. Interestingly, H3 relaxin appeared to mediate its matrix remodeling effects via RXFP1, which is naturally expressed in cardiac fibroblasts,[45] suggesting that both H2 and H3 relaxin act through RXFP1 to stimulate ECM remodeling. Furthermore, in all these studies, H2 and H3 relaxin demonstrated anti-fibrotic effects in all the diseased organs that they were applied to, but did not affect basal ECM and collagen remodeling in unaffected organs, demonstrating that relaxin is potent, but safe anti-fibrotic hormone, with rapid-occurring efficacy. In most cases, the therapeutic effects of relaxin were shown to be rapid and safe when administration resulted in circulating levels of 20-50 ng/ml, which is within the physiological range of serum relaxin observed in pregnant rodents.[1,2]

The Protective Effects of Endogenous Relaxin

Within many organs, there are a number of naturally occurring anti-fibrotic factors that are required to maintain homeostasis. Relaxin may be one such molecule. In the same way that hepatocyte growth factor and bone morphogenic protein-7 down-regulate transforming growth factor-beta1 (TGF β1) signaling by interfering with Smad signal transduction, relaxin also moderates fibrogenesis at several levels, by inhibiting the influence of several profibrotic factors; inhibiting fibroblast proliferation and/or differentiation; and stimulating MMP-induced matrix degradation.

The ability of endogenous relaxin to protect the rodent female reproductive tract and mammary apparatus (which was required for successful parturition and lactation) was first identified in a series of studies, which used neutralizing antibodies to rat relaxin.[47-49] Endogenous relaxin was confirmed to be required for maintaining the length of gestation, litter delivery and survival, which was attributed in part, to its ability to promote cervical softening and extensibility in addition to vaginal growth.[50] These observed changes in the late pregnant rat cervix and vagina were shown to be associated with a reduced density and organization of collagen fibre bundles and altered length of elastin fibres.[49,51]

To further our understanding of the physiological role of endogenous relaxin, scientists at the Howard Florey Institute generated a relaxin gene knockout (*Rln1-/-*) mouse.[52] This mouse was generated by means of gene targeting involving homologous recombination in embryonic stem (ES) cells, which was used to disrupt the mouse relaxin gene. The deleted fragment encoded 90 C-terminal amino acids of the 103 amino acids that make up the C peptide and 17 N-terminal amino acids of the 25 amino acids that make up the A chain, these regions were chosen as they are essential for the biological activity of relaxin. The deletion of the mouse relaxin gene in these mice caused the elimination of circulating relaxin,[52] while these mice still retained the mouse *Rln3* gene.

Pregnancy-Related Effects of Endogenous Relaxin

Late pregnant *Rln1-/-* mice underwent impaired development of the pubic symphysis, mammary gland, nipples and female reproductive organs during pregnancy[52] (Table 2), resulting in some level of impaired delivery and fetal survival and the inability of *Rln1-/-* mothers to lactate, consistent with the earlier studies which used neutralizing antibodies to rat relaxin.[47-51] The underlying pathology associated with these organs was attributed to a build up of collagen deposition and diminished collagen turnover, which would normally be mediated by relaxin in wild-type (*Rln1+/+*) mice.[53] *Rxfp1-/-* mice also demonstrated impaired nipple development during late pregnancy and were unable to feed their young,[54,55] suggesting that the effects of relaxin associated with pregnancy, were mediated via RXFP1 in vivo.

Developmental-Related Effects of Endogenous Relaxin

Male *Rln1-/-* mice also demonstrated underdeveloped reproductive tract development and inadequate growth of the prostate, testis and epididymis,[56] leading to impaired male fertility (Table 2). Again the underlying pathology associated with these findings involved increased collagen deposition in all these organs, in addition to decreased sperm maturation and increased cell apoptosis in ageing male *Rln1-/-* mice.[56] Similarly, male *Rln1-/-* mice underwent impaired spermatogenesis leading to azoospermia and a reduction in male fertility,[54] suggesting a role for the relaxin-RXFP1 interaction in the process of male reproductive tract development and function.

Table 2. Organs affected by increased collagen deposition (fibrosis) in the Rln1-/- mouse model

Organ Affected	Phenotype Observed	References
Pubic symphysis[a]	Inability to 'relax' and elongate	52,53
Cervix[a]	Impaired development and softening	52,53
Vagina[a]	Impaired development and growth	52,53
Mammary gland[a]	Impaired development; gross dilation of ducts	52,53
Nipple[a]	Impaired development; inability to enlarge and pass milk to new-born pups (lactate)	52,53
Prostate	Impaired growth and development of glandular epithelium	56
Testis	Impaired tubule development and regulation of spermatogenesis, leading to infertility	56
Epididymis	Impaired tubule development	56
Heart[b]	Increased atrial hypertrophy, left ventricular (LV) chamber stiffness and LV diastolic dysfunction	57
Kidney[b]	Increased weight, cortical thickening, interstitial and glomerular matrix and decreased renal function	59
Lung	Increased weight, alveolar congestion, bronchiole epithelium thickening and altered lung function	58
	[c]Increased airway hyperresponsiveness (AHR) and impaired regulation of MMPs	63
Skin	Increased dermal thickening	60

[a]Phenotype observed in late pregnant *Rln1-/-* mice and resulted in prolonged gestation and/or impaired delivery in some mice; [b]Phenotype only observed in male *Rln1-/-* mice; [c]Phenotype only observed in *Rln1-/-* mice subjected to a chronic ovalbumin-induced model of allergic airways disease. All other phenotypes observed in developing and/or ageing *Rln1-/-* mice.

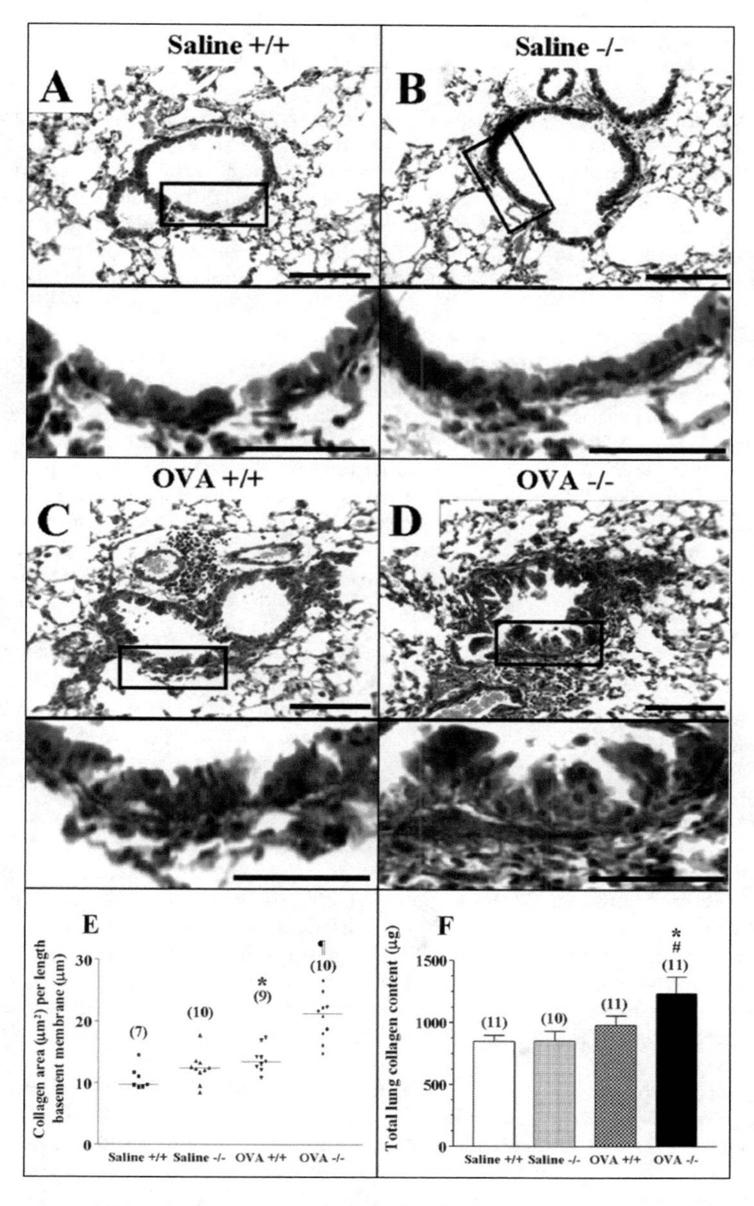

Figure 3. Lung collagen distribution and content in saline and OVA-treated mice. Masson trichrome stained lung tissue sections were examined to assess collagen distribution in Saline-treated *Rln1+/+* (A) and *Rln1-/-* (B) mice and OVA-treated *Rln1+/+* (C) and *Rln1-/-* (D) mice. Bar = 100µm in upper panels and 50µm in lower panels. Morphometric analysis was performed (E) and results expressed as the area of collagen per basement membrane length (n = 7-10 samples per group). *p < 0.02 when compared to saline-treated *Rln1+/+* mice; ¶p < 0.001 when compared to OVA-treated *Rln1+/+* mice. Hydroxyproline analysis of collagen content (F) was also performed on the four groups studied (n = 10-11 samples per group) and was expressed as the total collagen content per group. *p < 0.05 when compared to saline-treated *Rln1+/+* mice and #p < 0.05 when compared to saline-treated *Rln1-/-* mice. Reproduced from reference 63 with permission; ©2006, The Endocrine Society.

Age-Related Effects of Endogenous Relaxin

The effects of long-term relaxin-deficiency on ageing *Rln1-/-* mice have also been studied in several nonreproductive organs and have demonstrated an age-related progression of interstitial fibrosis in the heart,[57] lung,[58] kidney[59] and skin[60] of these animals, leading to organ damage and dysfunction (Table 2). The absence of endogenous relaxin in male *Rln1-/-* mice caused an-age dependent increase in left ventricle myocardial collagen content, atrial hypertrophy and impeded ventricular diastolic filling (due to increased ventricular chamber stiffness) from 9-months of age and onwards,[57] which was associated with increased ventricular type I collagen gene expression from 6-months of age and onwards. Interestingly this phenotype was not observed in female *Rln1-/-* mice, even up to 24-months of age.[57] Similarly, male but not female *Rln1-/-* mice underwent an age-related progression of interstitial renal fibrosis (involving elevated types I and III collagen) and increased glomerular matrix accumulation from 6-months of age and onwards, which resulted in increased cortical thickening and reduced renal function (creatinine clearance).[59] The lungs of both male and female *Rln1-/-* mice also underwent a progressive increase in airway/pulmonary fibrosis, which resulted in marked increases in organ weight, collagen accumulation, increased bronchiole epithelium thickening and altered lung function,[58] while both male and female *Rln1-/-* mice also demonstrated an age-dependent progression of dermal scarring and thickening, associated with increased types I and III collagen from 6-months of age and onwards.[60] In these organs (lung, skin) though, the severity of fibrosis was more prominent in male *Rln1-/-* mice.[58,60] These combined findings suggest that endogenous relaxin protects several reproductive and nonreproductive organs from the progression of collagen accumulation (fibrosis) and that gender plays a key role in influencing the effects of relaxin in these various organs.

Based on the gender-biased phenotypes that were observed in *Rln1-/-* mice, it was hypothesized that i) female *Rln1-/-* mice may have been protected from the progression of fibrosis by other sex steroids (such as estrogen), which may have compensated for the absence of relaxin, or ii) the presence of androgens in male *Rln1-/-* mice may have contributed to the severity of disease that they were associated with. In a recent study, the combined effects of estrogen and/or relaxin deficiency, in addition to estrogen replacement therapy (ERT) were analyzed in ovariectomized female *Rln1-/-* and age-matched wild-type mice.[61] Ovariectomy of both *Rln1-/-* and wild-type mice had no noticeable effects on cardiac or renal fibrosis, however, cardiac hypertrophy was significantly elevated in mice lacking both relaxin and estrogen.[61] In contrast, ovariectomy of *Rln1-/-* mice resulted in increased airway/pulmonary fibrosis, which resembled levels measured in age-matched male *Rln1-/-* by 12-months of age. ERT was able to reverse cardiac hypertrophy and airway/pulmonary fibrosis in these animals, suggesting that estrogen was able to contribute to relaxin's protective effect in inhibiting cardiac hypertrophy and airway/pulmonary fibrosis in female mice.[61] These findings may have important implications for postmenopausal women that are more prone to cardiovascular and pulmonary disease, due to the loss of estrogen and relaxin that they are associated with.

Importantly, we have also shown that the administration of recombinant H2 relaxin to older *Rln1-/-* mice with early and established stages of fibrosis can significantly reverse collagen accumulation in the heart,[37] lung,[58] kidney[59] and skin[60] over a 2-week treatment period. The rapid ability of H2 relaxin to decrease organ fibrosis resulted in the restoration of the structural integrity and function of these tissues to that observed in normal mice. Furthermore, H2 relaxin was most effective when administered to the early onset of disease, suggesting that its potency as an antifibrotic is inversely correlated to the severity of fibrosis. Interestingly, the lung,[55] kidney and heart (Samuel and Lekgabe, unpublished data) of *Rxfp1-/-* mice also undergo an age-related progression of collagen accumulation over time, which indicates that many of the anti-fibrotic or collagen-related actions of relaxin are mediated via the RXFP1 receptor in vivo. These findings have important implications for future research targeting prevention or reversal of fibrosis.

Induced Disease-Related Effects of Endogenous Relaxin

In more recent studies, the significance of endogenous relaxin in clinically relevant disease models has also been investigated using the *Rln1-/-* mouse. When subjected to a model of ovalbumin (OVA)-induced chronic allergic airways disease (AAD) over a 9-week period, which mimics several features of human asthma,[62] both OVA-treated *Rln1+/+* and *Rln1-/-* mice had increased airway inflammation, airway fibrosis and AHR, compared to that measured in saline-treated control animals (Fig. 3). However, airway fibrosis was significantly elevated in OVA-treated *Rln1-/-* mice, compared to that found in OVA-treated *Rln1+/+* mice (Fig. 3),[63] leading to increased AHR (Table 2), which correlated to either a reduction or failure to up-regulate MMP expression (in OVA-treated *Rln1-/-* animals).[63] These findings confirmed that the specific role of relaxin in this model involved its ability to regulate ECM and in particular, collagen turnover within the airways/lung.[63]

The relevance of endogenous relaxin has also been determined in a more rapidly progressive model of renal tubulointerstial fibrosis, induced by unilateral ureteric obstruction (UUO).[64,65] Inflammation, fibrosis, myofibroblast accumulation and MMP expression were all progressive elevated in the obstructed kidneys of *Rln1+/+* and *Rln1-/-* mice by 3-days post-UUO, however, only collagen and myofibroblast accumulation were significantly increased in *Rln1-/-* mice, compared to that measured in their *Rln1+/+* counterparts.[66] This increase in collagen and myofibroblast accumulation was reversed by the administration of H2 relaxin to *Rln1-/-* mice 3-days post-UUO, again confirming that relaxin played a specific role in regulating collagen in this model. Furthermore, no significant differences in inflammation nor MMP expression was observed between genotypes, suggesting that relaxin's ability to inhibit renal fibrosis was mediated through inhibition of renal myofibroblast accumulation in this model.[66] By 10-days post-UUO, the rapid progression of disease resulted in *Rln1+/+* mice having equivalent levels of collagen and myofibroblast accumulation to that measured in *Rln1-/-* mice, suggesting that endogenous relaxin only delays but does not prevent the progression of fibrosis in more rapidly occurring models of scarring.

These combined findings demonstrate the important protective role that endogenous relaxin plays in several organs to inhibit the progression of collagen accumulation, associated with the progression of fibrosis. Furthermore, they demonstrate that relaxin's ability to regulate collagen turnover has important implications during mammalian development and ageing, in addition to pregnancy.

Mechanism of Relaxin's Antifibrotic Properties

The ability of relaxin to downregulate collagen production and increase collagen-degradation is central to its physiological role. How it achieves this however, is less clear. Several in vitro studies have to date, provided the most promising insights into this process. As discussed above, H2 relaxin acts directly on TGF-β-stimulated fibroblasts from human[30-32,38] and rodent[34-37] origin to inhibit myofibroblast accumulation and collagen synthesis/secretion, while promoting MMP-induced collagen breakdown. Recent studies in human renal fibroblasts[33] suggested that these TGF-β-inhibitory effects of relaxin were mediated to a large extent by the Smad pathway. Phosphorylation of Smad2 and Smad3, formation of complexes with Smad4 and translocation of these complexes from the cytosol to the nucleus are key events in TGF-β-signaling.[67] Relaxin inhibited the phosphorylation and translocation of Smad2 to the nucleus, in the absence of any effects on phosphorylation of Smad3, c-Jun NH_2-terminal kinase 1/2, extracellular signal-regulated kinase (ERK), or p3MAP kinase.[33] These findings are of significance, given that Smads play an important role in the regulation of collagen accumulation. However, recent findings demonstrate that cross-talk among a variety of pathways is necessary for the maximal stimulation of collagen expression,[67] suggesting that relaxin, like many other ligands for G-protein coupled receptors, is capable of activating multiple signal transduction pathways.

This hypothesis has been confirmed by numerous other studies, which have demonstrated that relaxin acts on human endometrial cells to stimulate cAMP production and the activation of the p42/44 mitogen activated protein kinase (MAPK) and ERK pathways.[68] A similar activation of ERK and antagonism of endothelin-1 (a down-stream mediator of TGF-β) was observed when relaxin was administered to human umbilical vein endothelial cells and HeLa cells,[69] while relaxin has also been shown to mediate some of its effects via the PI-3 kinase pathway[70] and protein kinase C zeta translocation.[71] Further work is now required to determine which of these pathways is regulated by relaxin's matrix remodeling actions, on ECM-producing fibroblasts; and on this issue, recent studies have demonstrated that relaxin administration to rat ventricular fibroblasts[37] is associated with a transient rise in cAMP, which is no longer detected in these cardiac fibroblasts[37] nor human uterine fibroblasts[72] after ten minutes, suggesting that the actions of relaxin on fibroblasts are not primarily mediated via a Gs-protein mediated pathway.

Conclusions

Growing evidence suggests that relaxin is both an important endogenous regulator, which serves to protect ageing and diseased organs from the progression of fibrosis, in addition to being a potent, but safe antifibrotic agent. Relaxin is able to induce these antifibrotic effects via common and organ-specific mechanisms, primarily involving the inhibition of the ability of major profibrotic mediators (such as TGF-β, angiotensin II and their down-stream mediators) to promote fibroblast proliferation, differentiation and collagen/ECM production, in addition to an ability to stimulate MMP-induced collagen/ECM breakdown. It is important to note that while all these actions of relaxin are consistently identified in several organs, certain mechanisms of its antifibrotic actions may predominate, depending on the organ it is applied to. Identifying i) the nonreproductive sources of relaxin expression (as potential secondary sources), ii) the signaling mechanisms by which relaxin mediates its antifibrotic actions via fibroblasts and iii) which of these actions (which have been well studied in rodent models) applies to humans, will be essential to developing the clinical significance of relaxin. Importantly though, studies on relaxin and its receptor, RXFP1, offer us an exciting opportunity to better understand the pathophysiology of fibrosis in general.

Acknowledgements

The authors would like to thank Dr. Ross Bathgate for helpful suggestions regarding this chapter. E. D. Lekgabe is the recipient of a Jenny Ryan Scleroderma Foundation Postgraduate Scholarship, while I. Mookerjee is the recipient of a Howard Florey Institute Postgraduate Scholarship.

References

1. Bathgate RAD, Hsueh AJ, Sherwood OD. Physiology and Molecular Biology of the Relaxin Peptide Family. In: Knobil E, Neill JD, eds. Physiology of Reproduction. 3rd ed. San Diego: Elsevier 2006:679-770.
2. Sherwood OD. Relaxin's physiological roles and other diverse actions. Endocr Rev 2004; 25(2):205-234.
3. Hudson P, Haley J, John M et al. Structure of a genomic clone encoding biologically active human relaxin. Nature 1983; 301(5901):628-631.
4. Hudson P, John M, Crawford R et al. Relaxin gene expression in human ovaries and the predicted structure of a human preprorelaxin by analysis of cDNA clones. EMBO J 1984; 3(10):2333-2339.
5. Bathgate RA, Samuel CS, Burazin TC et al. Human relaxin gene 3 (H3) and the equivalent mouse relaxin (M3) gene. Novel members of the relaxin peptide family. J Biol Chem 2002; 277(2):1148-1157.
6. Hsu SY, Nakabayashi K, Nishi S et al. Activation of orphan receptors by the hormone relaxin. Science 2002; 295(5555):671-674.
7. Liu C, Eriste E, Sutton S et al. Identification of relaxin-3/INSL7 as an endogenous ligand for the orphan G-protein-coupled receptor GPCR135. J Biol Chem 2003; 278(50):50754-50764.
8. Liu C, Chen J, Sutton S et al. Identification of relaxin-3/INSL7 as a ligand for GPCR142. J Biol Chem 2003; 278(50):50765-50770.
9. Alberts B. Molecular biology of the cell. 3rd edition ed. New York: Garland Publishing; 1994.
10. Grodzinsky AJ, Frank EH, Kim Y-J et al. The role of specific macromolecules in cell-matrix interactions and in matrix function: physicochemical and mechanical mediators of chondrocyte biosymthesis. In: Comper WD, ed. Extracellular Matrix Vol. 2., Molecular Components and Interactions. Amsterdam, The Netherlands: Harwood Academic Publishers; 1996:310-334.
11. Lin CQ, Bissell MJ. Multi-faceted regulation of cell differentiation by extracellular matrix. FASEB J 1993; 7(9):737-743.

12. Rubin K, Gullberg D, Tomasini-Johansson B et al. Molecular recognition of the extracellular matrix by cell surface receptors. In: Kompa WD, ed. Extracellular Matrix Vol. 2., Molecular Components and Interactions. Amsterdam, The Netherlands: Harwood Academic Publishers; 1996:262-309.
13. Venstrom KA, Reichardt LF. Extracellular matrix. 2: Role of extracellular matrix molecules and their receptors in the nervous system. FASEB J 1993; 7(11):996-1003.
14. Bosman FT, Stamenkovic I. Functional structure and composition of the extracellular matrix. J Pathol 2003; 200(4):423-428.
15. Bateman JF, Lamande SR, Ramshaw JAM. Collagen Superfamily. In: Kompa WD, ed. Extracellular Matrix Vol. 2., Molecular Components and Interactions. Amsterdam, The Netherlands: Harwood Academic Publishers; 1996:22-67.
16. Kadler KE, Holmes DF, Trotter JA et al. Collagen fibril formation. Biochem J 1996; 316 (Pt 1):1-11.
17. Eddy AA. Molecular basis of renal fibrosis. Pediatr Nephrol 2000; 15(3-4):290-301.
18. Phan SH. The myofibroblast in pulmonary fibrosis. Chest 2002; 122(6 Suppl):286S-289S.
19. Samuel CS. Relaxin: antifibrotic properties and effects in models of disease. Clin Med Res 2005; 3(4):241-249.
20. Hisaw F. Experimental relaxation of the pubic ligament of the guinea pig. Proc Soc Exp Biol Med 1926; 23:661-663.
21. Weiss M, Nagelschmidt M, Struck H. Relaxin and collagen metabolism. Horm Metab Res 1979; 11(6):408-410.
22. Samuel CS, Butkus A, Coghlan JP et al. The effect of relaxin on collagen metabolism in the nonpregnant rat pubic symphysis: the influence of estrogen and progesterone in regulating relaxin activity. Endocrinology 1996; 137(9):3884-3890.
23. Wahl LM, Blandau RJ, Page RC. Effect of hormones on collagen metabolism and collagenase activity in the pubic symphysis ligament of the guinea pig. Endocrinology 1977; 100(2):571-579.
24. Downing SJ, Sherwood OD. The physiological role of relaxin in the pregnant rat. I. The influence of relaxin on parturition. Endocrinology 1985; 116(3):1200-1205.
25. Downing SJ, Sherwood OD. The physiological role of relaxin in the pregnant rat. II. The influence of relaxin on uterine contractile activity. Endocrinology 1985; 116(3):1206-1214.
26. Downing SJ, Sherwood OD. The physiological role of relaxin in the pregnant rat. III. The influence of relaxin on cervical extensibility. Endocrinology 1985; 116(3):1215-1220.
27. Downing SJ, Sherwood OD. The physiological role of relaxin in the pregnant rat. IV. The influence of relaxin on cervical collagen and glycosaminoglycans. Endocrinology 1986; 118(2):471-479.
28. Ferraiolo BL, Cronin M, Bakhit C et al. The pharmacokinetics and pharmacodynamics of a human relaxin in the mouse pubic symphysis bioassay. Endocrinology 1989; 125(6):2922-2926.
29. Samuel CS, Mookerjee I, Lekgabe ED. Actions of relaxin on nonreproductive tissues. Current Medicinal Chemistry-Immunology, Endocrine and Metabolic Agents 2005; 5(5):391-402.
30. Unemori EN, Amento EP. Relaxin modulates synthesis and secretion of procollagenase and collagen by human dermal fibroblasts. J Biol Chem 1990; 265(18):10681-10685.
31. Unemori EN, Pickford LB, Salles AL et al. Relaxin induces an extracellular matrix-degrading phenotype in human lung fibroblasts in vitro and inhibits lung fibrosis in a murine model in vivo. J Clin Invest 1996; 98(12):2739-2745.
32. Palejwala S, Stein DE, Weiss G et al. Relaxin positively regulates matrix metalloproteinase expression in human lower uterine segment fibroblasts using a tyrosine kinase signaling pathway. Endocrinology 2001; 142(8):3405-3413.
33. Heeg MH, Koziolek MJ, Vasko R et al. The antifibrotic effects of relaxin in human renal fibroblasts are mediated in part by inhibition of the Smad2 pathway. Kidney Int 2005; 68(1):96-109.
34. Williams EJ, Benyon RC, Trim N et al. Relaxin inhibits effective collagen deposition by cultured hepatic stellate cells and decreases rat liver fibrosis in vivo. Gut 2001; 49(4):577-583.
35. Bennett RG, Kharbanda KK, Tuma DJ. Inhibition of markers of hepatic stellate cell activation by the hormone relaxin. Biochem Pharmacol 2003; 66(5):867-874.
36. Masterson R, Hewitson TD, Kelynack K et al. Relaxin down-regulates renal fibroblast function and promotes matrix remodeling in vitro. Nephrol Dial Transplant 2004; 19(3):544-552.
37. Samuel CS, Unemori EN, Mookerjee I et al. Relaxin modulates cardiac fibroblast proliferation, differentiation and collagen production and reverses cardiac fibrosis in vivo. Endocrinology 2004; 145:4125-4133.
38. Unemori EN, Bauer EA, Amento EP. Relaxin alone and in conjunction with interferon-gamma decreases collagen synthesis by cultured human scleroderma fibroblasts. J Invest Dermatol 1992; 99(3):337-342.
39. Unemori EN, Beck LS, Lee WP et al. Human relaxin decreases collagen accumulation in vivo in two rodent models of fibrosis. J Invest Dermatol 1993; 101(3):280-285.
40. Garber SL, Mirochnik Y, Brecklin CS et al. Relaxin decreases renal interstitial fibrosis and slows progression of renal disease. Kidney Int 2001; 59(3):876-882.

41. Garber SL, Mirochnik Y, Brecklin C et al. Effect of relaxin in two models of renal mass reduction. Am J Nephrol 2003; 23(1):8-12.
42. McDonald GA, Sarkar P, Rennke H et al. Relaxin increases ubiquitin-dependent degradation of fibronectin in vitro and ameliorates renal fibrosis in vivo. Am J Physiol Renal Physiol 2003; 285(1):F59-67.
43. Lekgabe ED, Kiriazis H, Zhao C et al. Relaxin reverses cardiac and renal fibrosis in spontaneously hypertensive rats. Hypertension 2005; 46(2):412-418.
44. Zhang J, Qi YF, Geng B et al. Effect of relaxin on myocardial ischemia injury induced by isoproterenol. Peptides 2005; 26(9):1632-1639.
45. Bathgate RAD, Lin F, Hanson NF et al. Relaxin-3: Improved synthesis strategy and demonstration of its high affinity interaction with the relaxin receptor LGR7 both in vitro and in vivo. Biochemistry 2006; 45(3):1043-1053.
46. Silvertown JD, Walia JS, Summerlee AJ et al. Functional expression of mouse relaxin and mouse relaxin-3 in the lung from an Ebola virus glycoprotein-pseudotyped lentivirus via tracheal delivery. Endocrinology 2006; 147(8):3797-3808.
47. Hwang JJ, Shanks RD, Sherwood OD. Monoclonal antibodies specific for rat relaxin. IV. Passive immunization with monoclonal antibodies during the antepartum period reduces cervical growth and extensibility, disrupts birth and reduces pup survival in intact rats. Endocrinology 1989; 125(1):260-266.
48. Hwang JJ, Lee AB, Fields PA et al. Monoclonal antibodies specific for rat relaxin. V. Passive immunization with monoclonal antibodies throughout the second half of pregnancy disrupts development of the mammary apparatus and, hence, lactational performance in rats. Endocrinology 1991; 129(6):3034-3042.
49. Lee AB, Hwang JJ, Haab LM et al. Monoclonal antibodies specific for rat relaxin. VI. Passive immunization with monoclonal antibodies throughout the second half of pregnancy disrupts histological changes associated with cervical softening at parturition in rats. Endocrinology 1992; 130(4):2386-2391.
50. Zhao S, Kuenzi MJ, Sherwood OD. Monoclonal antibodies specific for rat relaxin. IX. Evidence that endogenous relaxin promotes growth of the vagina during the second half of pregnancy in rats. Endocrinology 1996; 137(2):425-430.
51. Zhao S, Sherwood OD. Monoclonal antibodies specific for rat relaxin. X. Endogenous relaxin induces changes in the histological characteristics of the rat vagina during the second half of pregnancy. Endocrinology 1998; 139(11):4726-4734.
52. Zhao L, Roche PJ, Gunnersen JM et al. Mice without a functional relaxin gene are unable to deliver milk to their pups. Endocrinology 1999; 140(1):445-453.
53. Zhao L, Samuel CS, Tregear GW et al. Collagen studies in late pregnant relaxin null mice. Biol Reprod 2000; 63(3):697-703.
54. Krajnc-Franken MA, van Disseldorp AJ, Koenders JE et al. Impaired nipple development and parturition in LGR7 knockout mice. Mol Cell Biol 2004; 24(2):687-696.
55. Kamat AA, Feng S, Bogatcheva NV et al. Genetic targeting of relaxin and Insl3 receptors in mice. Endocrinology 2004; 145(10):4712-4720.
56. Samuel CS, Tian H, Zhao L et al. Relaxin is a key mediator of prostate growth and male reproductive tract development. Lab Invest 2003; 83(7):1055-1067.
57. Du XJ, Samuel CS, Gao XM et al. Increased myocardial collagen and ventricular diastolic dysfunction in relaxin deficient mice: a gender-specific phenotype. Cardiovasc Res 2003; 57(2):395-404.
58. Samuel CS, Zhao C, Bathgate RA et al. Relaxin deficiency in mice is associated with an age-related progression of pulmonary fibrosis. FASEB J 2003; 17(1):121-123.
59. Samuel CS, Zhao C, Bond CP et al. Relaxin-1-deficient mice develop an age-related progression of renal fibrosis. Kidney Int 2004; 65(6):2054-2064.
60. Samuel CS, Zhao C, Yang Q et al. The relaxin gene knockout mouse: a model of progressive scleroderma. J Invest Dermatol 2005; 125(4):692-699.
61. Lekgabe ED, Royce SG, Hewitson TD et al. The effects of relaxin and estrogen deficiency on collagen deposition and hypertrophy of nonreproductive organs. Endocrinology 2006—in press.
62. Temelkovski J, Hogan SP, Shepherd DP et al. An improved murine model of asthma: selective airway inflammation, epithelial lesions and increased methacholine responsiveness following chronic exposure to aerosolised allergen. Thorax 1998; 53(10):849-856.
63. Mookerjee I, Solly NR, Royce SG et al. Endogenous relaxin regulates collagen deposition in an animal model of allergic airway disease. Endocrinology 2006; 147(2):754-761.
64. Klahr S, Morrissey J. Obstructive nephropathy and renal fibrosis. Am J Physiol Renal Physiol 2002; 283(5):F861-875.
65. Cochrane AL, Kett MM, Samuel CS et al. Renal structural and functional repair in a mouse model of reversal of ureteral obstruction. J Am Soc Nephrol 2005; 16(12):3623-3630.
66. Hewitson TD, Mookerjee I, Masterson R et al. Endogenous relaxin is a naturally occurring modulator of experimental renal tuubulointerstitial fibrosis. Endocrinology—in press.

67. Schnaper HW, Hayashida T, Hubchak SC et al. TGF-beta signal transduction and mesangial cell fibrogenesis. Am J Physiol Renal Physiol 2003; 284(2):F243-252.
68. Zhang Q, Liu SH, Erikson M et al. Relaxin activates the MAP kinase pathway in human endometrial stromal cells. J Cell Biochem 2002; 85(3):536-544.
69. Dschietzig T, Bartsch C, Richter C et al. Relaxin, a pregnancy hormone, is a functional endothelin-1 antagonist: attenuation of endothelin-1-mediated vasoconstriction by stimulation of endothelin type-B receptor expression via ERK-1/2 and nuclear factor-kappaB. Circ Res 2003; 92(1):32-40.
70. Nguyen BT, Yang L, Sanborn BM et al. Phosphoinositide 3-kinase activity is required for biphasic stimulation of cyclic adenosine 3', 5'-monophosphate by relaxin. Mol Endocrinol 2003; 17(6):1075-1084.
71. Nguyen BT, Dessauer CW. Relaxin stimulates protein kinase C zeta translocation: requirement for cyclic adenosine 3', 5'-monophosphate production. Mol Endocrinol 2005; 19(4):1012-1023.
72. Palejwala S, Stein D, Wojtczuk A et al. Demonstration of a relaxin receptor and relaxin-stimulated tyrosine phosphorylation in human lower uterine segment fibroblasts. Endocrinology 1998; 139(3):1208-1212.

Chapter 8

Relaxin-Like Ligand-Receptor Systems Are Autocrine/Paracrine Effectors in Tumor Cells and Modulate Cancer Progression and Tissue Invasiveness

Thomas Klonisch,* Joanna Bialek, Yvonne Radestock, Cuong Hoang-Vu and Sabine Hombach-Klonisch

Abstract

Relaxin and INSL3 are novel autocrine/paracrine insulin-like hormones in tumor biology. Both effectors can bind to and activate the leucine-rich G-protein coupled receptors LGR7 (relaxin receptor) or LGR8 (relaxin/INSL3 receptor). These relaxin-like ligand-receptor systems modulate cellular functions and activate signaling cascades in a tumor-specific context leading to changes in tumor cell proliferation, altered motility/migration and enhanced production/secretion of potent proteolytic enzymes. Matrix-metalloproteinases (MMP), tissue inhibitors of metalloproteinases (TIMP) and acid hydrolases such as cathepsins can facilitate tissue degradation and represent important proteolytic mediators of relaxin-like actions on tumor cell invasion and metastasis. This review presents recent new findings and emphasises the important functions of the relaxin/INSL3 ligand-receptor system as novel autocrine/paracrine effectors influencing tumor progression and tissue invasiveness.

Introduction

The relaxin-like peptide members relaxin and INSL3 have been detected in malignancies of the breast, gastrointestinal tract, thyroid gland, lung, colorectum and reproductive organs.[1-11] These ligands bind to the Leucin-rich G-protein-coupled Receptors LGR7 (relaxin receptor) and LGR8 (INSL3/ relaxin receptor). Relaxin, INSL3 and their cognate receptors can be found in carcinoma cells and neighboring stromal cells indicating potential autocrine and paracrine functions for these ligand-receptor systems in a diverse spectrum of tumor tissues. The functions of relaxin and INSL3 in normal and cancer tissues can be modulated by multiple endocrine signaling pathways in an organ- and tissue-specific manner as a result of complex interactions with nuclear steroid receptors activated by glucocorticoids/ progestins, androgens, estrogens and retinoic acids.

The growing list of tumor tissues with a potential involvement of the relaxin-like ligand-receptor system, includes carcinoma of the mammary and thyroid gland and malignancies of the male and female reproductive tract. Carcinoma cell lines of mammary, prostate, thyroid, trophoblast and myelomonocytic origin and cellular transfectants producing and secreting bioactive H2-relaxin

*Corresponding Author: Thomas Klonisch—Department of Human Anatomy and Cell Science, Faculty of Medicine, University of Manitoba, 130 Basic Medical Science Bldg., 730 William Avenue, Winnipeg, Manitoba, Canada, R3E 0W3. Email: klonisch@cc.umanitoba.ca

Relaxin and Related Peptides, edited by Alexander I. Agoulnik.

(RLN2) and INSL3 have served as valuable tools for in vitro investigations: 1) to elucidate the cellular functions and molecular mechanisms of both relaxin and INSL3 and 2) to identify novel relaxin-like target molecules in carcinoma cells.

The myelomonocytic cell lines THP-1 and U-937 were the first human cancer cell lines identified as being relaxin-responsive and both human leukaemia cell lines continue to serve as models for the study of relaxin-mediated signaling events.[11-14] Mice with a homozygous deletion of the genes for relaxin and/or INSL3 (single/ double knockout mutants) or their cognate receptors Lgr7 and Lgr8 do not per se present with an enhanced tumor incidence. Homozygous deletion of the Insl3 or Lgr8 gene, however, dramatically impairs the ability of the testis to descent into the scrotum.[15-18] Whereas LGR7-expressing human testis may be a target organ for the systemic actions of relaxin, adult testicular Leydig cells produce significant amounts of the Leydig cell marker INSL3 to bind to LGR8 INSL3/relaxin receptors present in the human testis.[19,20] It has been reported that mice with their Lgr7 gene deleted display a transitory disruption of spermatogenesis and reduced fertility, likely as a result of increased apoptosis of stage 12 meiotic cells observed exclusively in F1/2 generations of Lgr7-deficient mice.[21] Leydig cell-derived INSL3 appears to reduce apoptosis of Lgr8-positive testicular germ cells in the normal mouse testis.[22] Spermatogonia have recently been identified to possess stem cell-like capabilities and testicular stem cell survival may in part depend on a functional INSL3-LGR8 ligand-receptor system.[23] A role for the INSL3 and/or LGR8 in human testicular tumors is currently unknown but the expression of INSL3 in human testicular *Leydig cell neoplasia* was found markedly reduced (Fig. 1).[10] This downregulation in INSL3 production in human Leydig cell adenoma of the adult testis could reflect the appearance of Leydig tumor cells with a less differentiated prepubertal phenotype.[10] Rodent Leydig cell lines derived from testicular Leydig cell tumors also express very low Insl3 mRNA levels, much like prepubertal mouse testicular Leydig cells.[24,25] The roe deer as a seasonal breeder is another example for a differentiation-dependent expression of INSL3 in testicular Leydig cells. Here, INSL3 production by Leydig cells is low during the nonreproductive months but is up-regulated strongly in an LH-independent manner prior to the brief rutting period.[26] In the human testis, increased expression of estrogen receptors (ER) coincides with a malignant transition of Leydig cells.[27] The enhanced ER presence and the ability of Leydig cells to synthezise and secrete estrogens is suspected to activate an autocrine self-enhancing proliferative cycle in neoplastic Leydig cells.[27-32] Estrogens

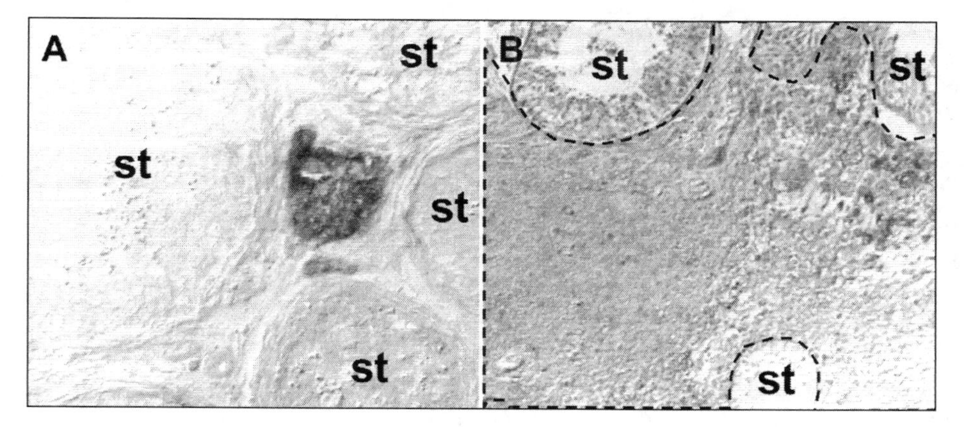

Figure 1. In-situ hybridization analysis for *INSL3* in a normal human testis (A) and in a testis with Leydig cell adenoma (B). Strong expression of *INSL3* transcripts was observed in normal testicular Leydig cells located in the intertubular space between seminiferous tubules (st). By contrast, in Leydig cell adenoma (B; dotted area) *INSL3* transcripts were only detected in the periphery but expression was weak to absent in the centre of this testicular tumor (B). Identical results were obtained when a specific antiserum against human INSL3 was employed.[10]

have been identified as inhibitors of Insl3 gene activity and the same mechanism could explain an estrogen-mediated downregulation of INSL3 production in human Leydig cell tumors.[7,10,33]

In this review, we summarize the current knowledge of the expression and cellular actions of the relaxin-like ligand-receptor system with a particular emphasis on the role of relaxin and INSL3 in prostate, thyroid and breast cancer tissues and recent findings in corresponding carcinoma cell lines.

Prostate Cancer

Despite all efforts towards earlier detection and aggressive treatment the mortality rate for prostate carcinoma, the most common cancer in men, is steadily increasing.[34] In the human and armadillo, relaxin is known to be expressed in the epithelial cell layer of the prostate gland[8,35,36] and more recently *LGR7* transcripts were demonstrated in normal human prostate tissues.[20] The epithelium of benign prostate hyperplasia (BPH), in particular the basal prostate epithelial cells and carcinoma cells from prostate cancer at all pTNM stages were identified as a source and potential target tissue of the INSL3-LGR8 ligand-receptor system (Fig. 2).[35]

The physiological role of any of the members of the relaxin-like ligand-receptor system in the human or mouse prostate is currently unknown. A report on relaxin-knockout (KO) mice having a significantly smaller prostate gland more densely packed with collagen fibres as compared to similar tissue obtained from control mice[37] could not be confirmed in relaxin receptor Lgr7-KO, Lgr8-KO and double Lgr7-/Lgr8-double deficient mice,[38] with no information available as yet from Insl3-KO mice.

Of three human prostate adenocarcinoma cell lines investigated, both androgen-insensitive cell lines, PC-3 and DU-145, expressed *INSL3* transcripts and, similar to the alternatively-spliced larger human relaxin transcripts, DU-145 also expressed a larger *INSL3* splice variant first described in the human anaplastic thyroid carcinoma cell line 8505C.[7,39] The androgen-sensitive and *RLN1* and *RLN2* relaxin-positive LNCaP was devoid of *INSL3* isoforms but expressed *LGR8* transcripts and, thus, may be regarded INSL3-sensitive.[39,40] Human PC-3 cells co-expressed transcripts encoding *INSL3* and both the *LGR7* and *LGR8* receptors.[35] PC-3 and DU-145 also co-expressed *RLN2*[39] indicating that these cancer cell lines may possess a potentially functional autocrine/paracrine RLN2-LGR7 system and, in the case of PC-3, also a functional INSL3-LGR8 ligand-receptor signaling system. Further evidence for a local paracrine effect of INSL3 and relaxin produced by the BPH epithelium or prostate carcinoma cells on neighboring prostate interstitial cells derives from the fact that the human prostate interstitial stromal cell line hPCP[41] also expresses *LGR7* and *LGR8* transcripts (T. Klonisch, personal communication). In support for an autocrine/paracrine mode

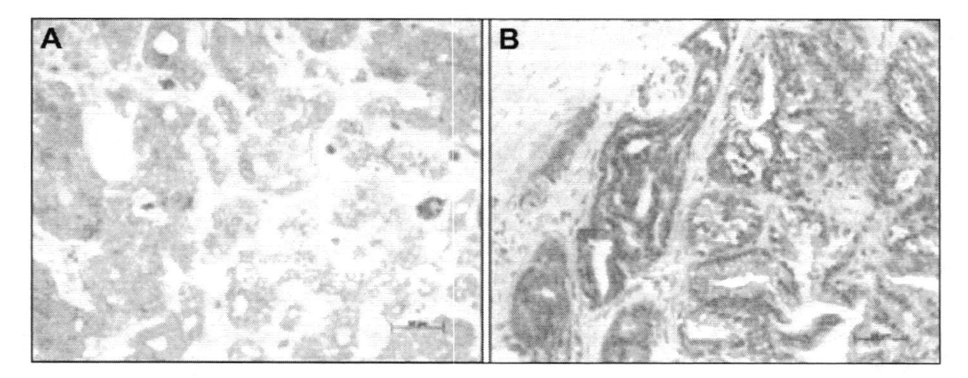

Figure 2. The presence of immunoreactive INSL3 (A) and transcripts for *LGR8* (B) was detected in human prostate carcinoma tissues. The role of this INSL3-LGR8 ligand-receptor system during prostate carcinogenesis is currently unknown and an area of ongoing investigations. Reproduced with permission from Klonisch et al, Int J Oncol 2005.

of action, PC-3 prostate carcinoma cells overexpressing RLN2 were shown to enhance prostate xenograph growth and angiogenesis.[42] We found that INSL3 enhanced the cellular motility of PC-3 cells in a concentration-dependent manner, with higher INSL3 concentrations being equally effective to the known prostate carcinoma cell promigratory growth factor EGF.[35] The EGF-ErbB system is kown as an important mediator of enhanced tissue-invasiveness in prostate cancer.[43-45] However, this membrane-anchored tyrosine kinase receptor system appeared not to be a major contributor to the INSL3-mediated enhanced in vitro motility/migration of human prostate carcinoma cells. Instead, the promigratory effect of human INSL3 coincided with a downregulation of the gene activities of the ErbB receptors, particularly ErbB 1-3 and EGF ligand in PC-3.[35] We also identified all-trans-retinoic acid (RA) to be a novel selective activator of *LGR8* gene activity in PC-3 prostate carcinoma cells, with no effect on the transcription of the INSL3 gene.[35] This would suggest specific DNA binding sites for nuclear retinoic acid receptors (RAR-α, -β, -γ) to be present in the promoter region of the human *LGR8* gene.[46] While these results would suggest an intricate relationship between abnormal retinoid nutritional states,[47,48] altered RAR expression[49-51] and LGR8-mediated signaling in human prostate cancer,[35] the functional relevance of these in vitro findings awaits further clarification in the normal or malign human prostate gland.

Novel relaxin-like splice transcripts are present and numerous factors regulate the expression of the relaxin-like ligand receptor system in normal and malign human prostate cells. Present in the human prostate epithelium are *RLN1* and *RLN2* and alternatively spliced *RLN1* and *RLN2* transcripts containing a 101 bp splicing-in gene fragment derived of an additional exon.[39] Androgen-independent PC-3 and DU-145 human prostate adenocarcinoma cell lines exclusively express *RLN2*, whereas LNCaP.FGC contain transcripts for both *RLN1* and *RLN2*.[52] *RLN1*- and *RLN2*-chloramphenicol acetyltransferase (CAT) reporter constructs revealed the *RLN1*-CAT promoter to be substantially less active than the corresponding *RLN2*-CAT construct when transfected into the androgen-responsive human adenocarcinoma cell line LNCaP.FGC. Differences in the TATA boxes between the *RLN1* and *RLN2* gene promoters were thought to attribute to these differences in promoter activity.[52] Similar to LNCaP.FGC, JAR human choriocarcinoma cells co-express *RLN1* and *RLN2*.[51,52] Relaxin promoter reporter assays suggested a differential role for progesterone receptor (PR) and glucocorticoid receptor (GR) in the regulation of *RLN1* and *RLN2* expression in JAR.[53] Medroxyprogesterone acetate (MPA) caused *RLN2* gene activity to increase and the glucocorticoid dexamethasone significantly upregulated both *RLN1* and *RLN2* expression. The antiprogestin RU486 blocked the binding of GR to glucocorticoid response elements (GRE) within the 5'-region of the relaxin genes but enhanced binding of PR to these GREs.[53] Although the role of PR and GR for the regulation of relaxin genes in the normal and neoplastic human prostate is currently unknown, both PR and GR have been described in human prostate carcinoma.[54,55] Highest expression of human PR has been reported in metastatic and androgen-insensitive prostate tumors and elevated expression of GR has been correlated with a much reduced level of AR in human prostate cancer tissues.[53,54] Increased production of immunoreactive relaxin was reported in human prostate tissues from patients receiving anti-androgenic treatment and in prostate carcinoma cells of bone metastases.[40] Although androgen withdrawal was recently shown to increase *RLN2* gene activity in the androgen-responsive LNCaP human prostate carcinoma cell line,[40] this was not followed by an upregulation of RLN2 protein production.[56] LNCaP transfectants harbouring the gain-of-function p53 mutant R273H ($p53^{R273H}$) mutation and to a lesser extend p53 mutations of residues G245S, R248W and R273C, revealed an upregulation of *RLN2* gene transcription. However, only LNCaP-$p53^{R273H}$ transfectants produced and secreted immunoreactive relaxin which acted as an effector of an autocrine signaling survival mechanism involving LGR7.[56] The $p53^{R273H}$ mutation is frequent in androgen-independent prostate carcinoma patients[57] and $p53^{R273H}$ was shown to directly bind to the RLN2 promoter, thus, a) linking the $p53^{R273H}$ mutation with the downstream effector function of the RLN2-LGR7 system and b) identifying RLN2 as an important mediator of androgen-independent growth of LNCaP-$p53^{R273H}$ mutants.[56] Furthermore, Vinall et al[56] demonstrated for the first time that relaxin could activate AR signaling pathways in both androgen-dependent LNCaP and androgen-independent DU145 in the absence of androgens. These findings are intriguing as RLN2 was also shown to directly bind to and activate the glucocorticoid receptor (GR) signaling pathway.[58,59]

Thyroid Cancer

Thyroid cancer is derived from follicular thyrocytes or parafollicular C-cells and is one of the most frequent endocrine-related cancers in men.[60,61] Relaxin may have a physiological role in the thyroid gland. Earlier reports of crude relaxin preparations causing an increase in thyroid weights, radioactive iodine uptake and protein-bound iodination in rats remain controversial.[62-64] Besides the presence of relaxin/INSL3 receptor *LGR8* mRNA,[7] relaxin receptor LGR7 mRNA and protein were recently identified in human normal and malignant thyrocytes, identifying the human thyroid gland and thyroid carcinoma as novel targets of systemic relaxin.[20,65] Furthermore, relaxin is a product of neoplastic but not normal human thyrocytes and acts as a novel autocrine/paracrine effector hormone in human thyroid carcinoma.[65] INSL3 and a novel INSL3 splice variant were demonstrated in hyperplastic thyrocytes of Graves' disease and in human thyroid carcinoma tissues suggesting distinct functional differences between relaxin and INSL3 in the human thyroid (Fig. 3).[7]

Both proforms of RLN2 and INSL3 were found to be the main product in human thyroid carcinoma tissues and cell lines.[7,65] This apparent lack in complete processing of relaxin-like members by human normal and malignant thyrocytes is known not to affect relaxin-like ligand binding and receptor activation.[66-68]

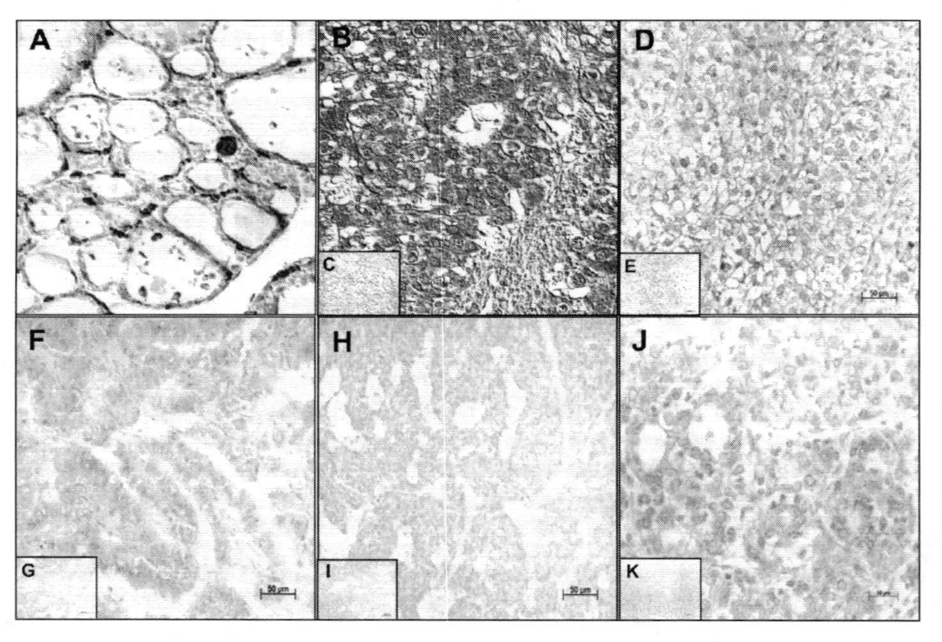

Figure 3. Immunodetection of INSL3 (A-E), relaxin (F-I) and LGR7 (J, K) in human normal (A) and neoplastic thyroid tissues (B-K). Immunoreactive INSL3 was detected in hyperplastic and neoplastic thyrocytes as shown for undifferentiated thyroid carcinoma (UTC; B). INSL3 was absent in the normal thyroid gland (A). Relaxin was exclusively detected in UTC (D), papillary (PTC; F) and follicular thyroid carcinoma (FTC; H) but was absent in the normal and hyperplastic thyroid gland. Relaxin receptor LGR7 was strongly expressed in thyroid carcinoma as shown for FTC (J). Tissue sections treated with non-immune rabbit serum were devoid of specific immunostaining (C, E, G, I, K). Reproduced with permission from Hombach-Klonisch et al, Am J Pathol 2006.

Estrogens may modulate the expression of the INSL3-LGR8 ligand receptor system in malignant human thyrocytes. Diethylstilbestrol treatment was reported to cause a down-regulation of *INSL3* but an upregulation in *LGR8* transcriptional gene activity in the human undifferentiated anaplastic thyroid carcinoma cell line 8505C.[7] It is tempting to suggest a potential intra-thyroidal signaling network involving both the INSL3-LGR8 and the estrogen-ER ligand-receptor system.

Relaxin-like peptides may not just act as effector hormones for thyrocytes but also for thyroidal calcitonin-producing C-cells within the thyroid gland. Both RLN2 and INSL3 were identified as specific products of human hyperplastic parafollicular thyroid C-cells and medullary thyroid carcinoma (MTC) derived from neoplastic C-cells.[69] Both relaxin-like hormones were not detected in normal, calcitonin-positive C-cells of benign goiter tissues. The detection of *LGR7* and *LGR8* transcripts in human MTC tissues and human and mouse MTC cell lines suggests a potentially functional relaxin-like ligand-receptor signaling system in this thyroid tumor entity which accounts for 5-10% of all thyroid malignancies. *RLN2* was a common transcript in primary human MTC tissues, while *INSL3* was more frequently observed in MTC metastases.[69] *Relaxin* and *INSL3* expression profiles were independent of the pTNM classification status of the MTC investigated. *LGR7* and *LGR8* transcripts were present in all MTC tissues investigated.[69] Despite calcitonin production being down-regulated in MTC of multiple endocrine neoplasia MEN2a and MTC metastases, expression of relaxin-like members was unaltered, identifying relaxin and INSL3 as potential novel, calcitonin-independent tumor markers in MTC. A role for relaxin and INSL3 as autocrine/paracrine thyroid factors is also suggested in the *RLN2*- and *INSL3*-expressing human MTC-derived cell line TT and the *Insl3*-expressing mouse MTC-M cell line which both express *LGR7* and *LGR8* transcripts.[7,65,69] Currently, however, the functional significance of relaxin and INSL3 in MTC remains elusive.

Recent studies employing stable transfectants of human follicular thyroid carcinoma cell lines overexpressing and secreting bioactive proRLN2 (FTC-133 and FTC-238) and human proINSL3 (FTC-133) have shed light on the functional roles of RLN2 and INSL3 in human thyroid carcinoma cells. FTC-133 and FTC-238 express bioactive LGR7 and LGR8.[65] Microarray analysis of FTC-133-Rln and FTC-133-Insl3 clones revealed 432 and 522 genes, respectively, associated with specific functions and disease entities. Cancer-related genes formed the largest subset of genes induced by relaxin (248 genes) and INSL3 (388 genes = cancer: 307 genes + tumor morphology: 81 genes) in FTC-133 (Fig. 4).

The effect of both ligand-receptor systems on tumor cell growth appears to depend on (i) the specific tumor cells under investigation and (ii) the environment these tumor cells grow in. For example, neither relaxin nor INSL3 acted as a growth factor on thyroid carcinoma cells in culture dishes. Similarly, the human endometrial carcinoma cell lines HEC-1B and KLE expressed *LGR7* transcripts but relaxin treatment did not stimulate proliferation in-vitro.[70] When nude mice were subcutaneously injected with FTC-133 secreting bioactive proRLN2 and to a lesser extend also human proINSL3 (T. Klonisch, personal communication), they formed rapidly growing tumors (Fig. 5).

Both recombinant RLN2 and secreted proRLN2 profoundly accelerated the cellular in vitro motility of the two human thyroid carcinoma cell lines FTC-133 and FTC-238 and the corresponding RLN2 transfectants in an autocrine/paracrine LGR7-dependent manner.[65] This ability to promote tumor cell motility is characteristic for both relaxin and INSL3. Relaxin is known to promote the motility of human and canine mammary carcinoma cell lines (human: MCF-7 and SK-BR3; canine: CF33.Mt)[66,71] as well as noncarcinogenic bronchial epithelial cells and inflammatory cells infiltrating wounds.[72,73] INSL3 acts as a motility-enhancing factor on PC-3 human prostate carcinoma cells[35] and human neoplastic thyrocytes (T. Klonisch; personal communication). Apart from MMPs and their tissue inhibitors of metalloproteinases (TIMPs) known to facilitate relaxin's effects on migration and ECM invasion of mammary and endometrial carcinoma cells,[70,71,74-76] the highly potent proteolytic lysosomal acid hydrolases of the cathepsin (cath) family were recently identified as novel relaxin target enzymes in thyroid

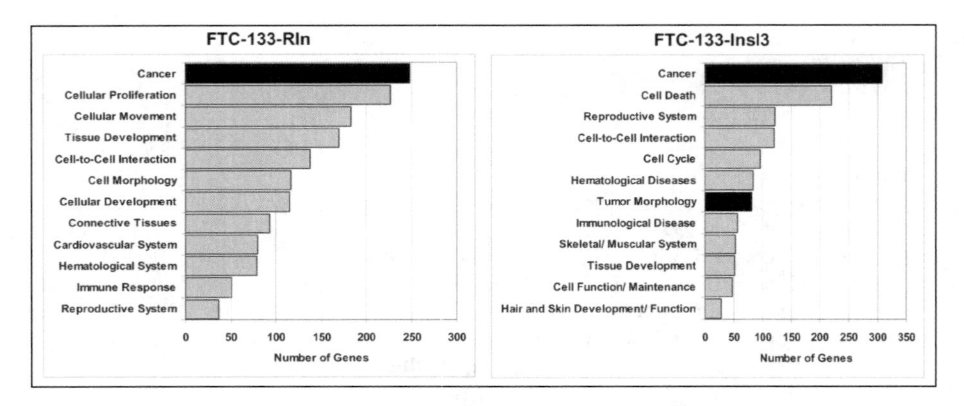

Figure 4. Microarray analysis of Affymetrix U-133 chips of stable FTC-133 transfectants overexpressing human proRLN2 and proINSL3 revealed a significantly altered transcription of 432 and 522 genes, respectively. When these genes were grouped according to function, disease entities and statistical significance using the Ingenuity Systems software, cancer-associated genes comprised the largest of the 12 most significant function/disease groups in both FTC-133-Rln and FTC-133-Insl3 transfectants.

carcinoma cells.[65] FTC-133-RLN2 clones displayed a significantly increased elastinolytic activity.[65] Numerous pathologies such as inflammatory processes,[77] Alzheimer's disease[78,79] and metastasis of tumor cells[80-83] contain a significant involvement of cathepsin proteases. Cath-D and cath-L produced by FTC-133-RLN2 transfectants are active in human thyroid carcinoma tissues.[83-85] The upregulation in the production and secretion of cath-L forms in RLN2 transfectants correlated with an enhanced elastinolytic activity and the ability to invade elastin matrices. As a powerful protease promoting migration and basement membrane degradation by tumor cells,[86-88] cath-L also confers to endothelial progenitor cells the proteolytic, pro-migratory capacity essential for

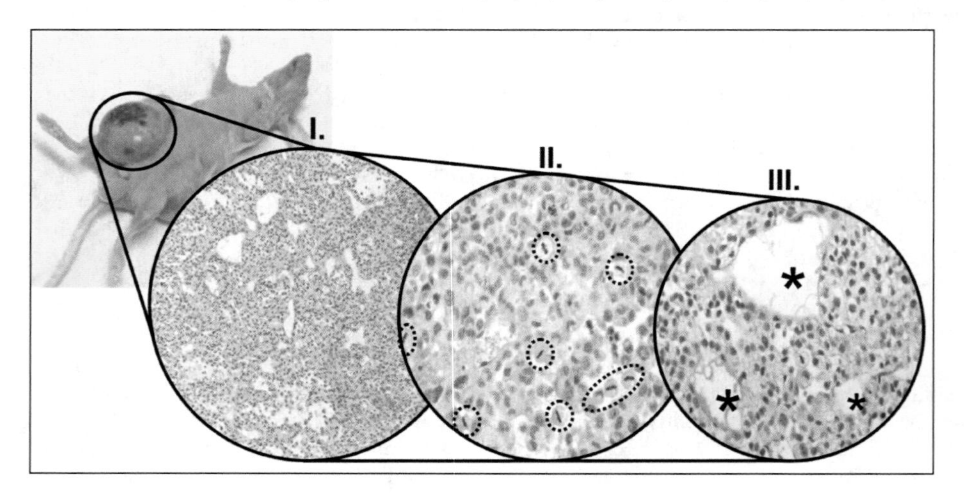

Figure 5. FTC-133 overexpressing and secreting bioactive proRLN2 created large and fast-growing subcutaneous tumors in nude mice within three weeks after inoculation. These tumors were highly vascularized (I.+ III.; stars in III. indicate newly formed blood vessels) and contained numerous mitotically active human thyroid carcinoma transfectants as indicated by the frequent metaphases (II. circled in dotted lines).

neovascularization in vivo.[89] Relaxin affected the cytosolic distribution of cath-L in RLN2 thyroid carcinoma transfectants with a polar to perinuclear distribution of (pro)cath-L in these RLN2 clones as opposed to an even granular cytoplasmic distribution in the EGFP controls. The exact role for relaxin in this process is unclear but it may involve changes in the endosome to lysosome routing of mannose-6-phosphate-tagged cath-L since mannose-6-phosphate receptors[90] displayed a cytoplasmic distribution similar to that detected for procath-L in the RLN2 transfectants. Relaxin appeared to specifically affect trafficking of cath-L containing vesicles since no change in the cytoplasmic distribution for vesicles containing cath-D was observed in the RLN2 transfectants.[65] In thyroid carcinoma tissues, cath-D concentrations are significantly higher and correlate with tumor size and stage.[85] RLN2 transfectants had significantly increased levels of procath-D and, like in human breast cancer, procath-D was the major secreted glycoprotein in all RLN2 transfectants investigated. Abundant production of relaxin in thyroid neoplasia implicates relaxin as a potential novel local regulator of procath-D production in human thyroid carcinoma. In contrast to human breast cancer, estrogens are not regarded as regulators of cath-D production in thyroid carcinoma tissues.[84,85] By up-regulating the production and secretion of (pro)cath-D and -L, relaxin may not only enhance the local proteolytic activity and tumor cell invasiveness but will also promote tumor angiogenic capacity and survival of human thyroid carcinoma cells.[91-94]

Cancer of Female Organs with Reproductive Functions

A majority of human endometrial carcinoma (EC) tissues and the two relaxin-expressing EC cell lines, HEC-1B and KLE, investigated were shown to express bioactive LGR7. Strong RLN expression was associated with (i) high-stage (III/IV) EC (67%) compared to low-stage (I/II) EC cases (37%), (ii) high-grade and depth of myometrial EC invasion and (iii) a significantly shorter overall survival compared to weak or moderate RLN expression.[70] Exogenous RLN2 stimulation caused significant LGR7-mediated increase in migration and invasion in HEC-1B and KLE. This correlated with increased levels of activated MMP-2 in KLE cells and activated MMP-9 in HEC-1B cells.[70]

Breast cancer is the most common cancer in women in the U.S., with an estimated 213,000 new cases diagnosed in the U.S. in 2005.[95] Relaxin-like peptide hormones and their cognate receptors are present in breast cancer. Immunoreactive relaxin is present in lobular and ductal mammary epithelium and in myoepithelial cells of normal human breast tissue derived from prepubertal, cyclic, gestational, lactational and postmenopausal women.[1-3] The lobular and ductal epithelium of normal postpubertal human breast tissues is also a source of INSL3 production in the breast.[4] Their expression within the mammary epithelium suggests an as yet largely unknown role for relaxin and INSL3 in the normal physiology of the human breast. Relaxin may act as a paracrine factor in the human mammary gland. Women without detectable serum levels of relaxin were found to contain immunoreactive relaxin in postpartum milk, in breast cyst fluid of patients suffering from mammary dysplasia and in the plasma of a woman with giant fibroadema of the breast.[1,96]

Relaxin does not act as an essential modulator of the secretory capacity of the mammary parenchyma in gilts.[97] In the pig and ginea pig, relaxin is expressed in alveolar epithelia and functions in concert with estrogen and progesterone to promote growth and differentiation of the mammary parenchyma during development and gestation.[97-101] By contrast, in rodents relaxin plays a minor role in the growth, differentiation and lactational performance of mammary gland tissue during pregnancy.[102] Relaxin has been shown to be essential for nipple development and nipple growth during gestation, affecting the composition of the extracellular matrix, the number of blood vessels and the size of lactiferous ducts within the nipples.[103,104] Impaired nipple development was described in both homozygous relaxin KO and LGR7 KO mice resulting in the offspring dying of starvation because of the inability of pubs to suckle.[21,104]

Currently, the localization of relaxin binding sites in the mammary gland and nipple remains contradictory. Min and Sherwood (1996) described relaxin binding to epithelial cells of alveola and lactiferous ducts and to smooth muscle cells and blood vessels of the mammary and nipple tissue using a biotinylated porcine relaxin tissue binding assay.[100,105] Ivell et al (2003) employed a rabbit polyclonal antibody generated against peptides of the ectodomain and the third intracellular loop of the LGR7 relaxin-receptor and exclusively immunolocalized LGR7 to the stromal compartment adjacent to the glandular epithelium in human healthy breast tissues and a mammary neoplasia.[106] These differences in the detection of relaxin-binding sites may be partially explained by the different methods used in the two studies and the fact that biotinylated relaxin may have crossreacted with LGR8 relaxin/INSL3 receptor.[20]

Relaxin-like peptides appear to act as autocrine/paracrine factors in human breast cancer tissues. Relaxin appears not to be a cancer-protective factor in the breast of pregnant rats.[107] *LGR7* and *LGR8* transcripts are expressed in malignant human breast tissues and in human mammary tumor cell lines.[108] Neoplastic epithelial cells of ductal, lobular, intraductal and infiltrating lobular breast tumors revealed enhanced cytoplasmic immunostaining for relaxin and expressed higher gene copy numbers for *RLN1* and *RLN2*.[1] *RLN2* expression is strongly associated with breast cancer and was detected in 100% of neoplastic breast tissues, but in only a small percentage in nonneoplastic, normal tissues. By contrast, *RLN1* transcripts were found in 75% of neoplastic tissues but in only 12.5% of the normal tissues suggesting that both human relaxin isoforms may perform different and so far unidentified functions in the mammary gland.

Porcine relaxin has a dose-dependent biphasic effect by promoting growth or differentiation of MCF-7 human mammary adenocarcinoma cells[109,110] and relaxin-treated MCF-7 cells in co-culture with myoepithelial cells differentiated towards an epithelial phenotype resembling normal mammary epithelial cells with an increase in cytoplasmic organelles, apical microvilli, intercellular junctions and microtubules.[111,112] MCF-7 transplants in nude mice treated with 17-beta-estradiol to promote tumor growth revealed that systemically applied relaxin (10 μg/day for 19 consecutive days) promoted differentiation of MCF-7 breast cancer cells into either epithelial-like or myoepithelial-like cells.[113] This dual differentiation potential is typical for and a physiological characteristic of pluripotent mammary epithelial precursor cells located within the outermost layer of the terminal end bud of the mammary ducts[114] suggesting that relaxin may induce epithelial differentiation in a stem cell population of MCF-7.[110] Whether these differentiation effects of relaxin on ER-positive MCF-7 are ER-alpha dependent, as shown for the estrogenic effects elicited by relaxin in the rat uterus,[115,116] is currently unknown. Human mammary gland-derived progenitor stem cells are reported to be devoid of ER-alpha and ER-beta and appear to be able to proliferate in an estrogen-independent manner.[117] An effect of relaxin on angiogenesis has not been investigated in the mammary gland but an induction of VEGF by relaxin has been described during wound healing.[73]

Detectable serum relaxin levels are only observed at the end of the first trimester in pregnant women and reflect increased production in the decidua and placenta, but are not detectable in cycling and lactating women.[118,119] Metastatic breast cancer can elevate relaxin serum levels in nonpregnant women raising the question as to whether relaxin secretion by mammary carcinoma cells promotes invasiveness in breast cancer patients.[120] Human and canine breast cancer cell lines have been used to test the influence of relaxin on cancer cell migration. Using matrigel-coated membranes, Binder et al (2002) reported enhanced in vitro invasiveness in the human breast cancer cell lines SK-BR3 and MCF-7 when treated with relaxin.[71] In both cell lines, relaxin increased the secretion of the matrix-metalloproteinases MMP-2, MMP-7 and MMP-9. Enhanced invasiveness was blocked in the presence of the broad spectrum MMP-inhibitor FN-439.[71] This may suggest a causal link between the increased expression of these matrix-degrading enzymes and relaxin-induced increased invasiveness of mammary carcinoma cells. MMP-9 has previously been shown to increase invasiveness in MDA-MB-231 human breast cancer cells by promoting the ability to degrade the extracellular matrix (ECM) and transverse the basement membrane.[121,122] Exogenously-delivered recombinant human relaxin at doses greater than 250 ng/ml or adenoviral-mediated expression of human recombinant prorelaxin induced an invasive phenotype in the canine mammary cancer cell line, CF33.Mt,[66] confirming that relaxin, like in human thyroid carcinoma cells,[65] is a potential pro-invasive hormone in breast cancer cells.

Preliminary studies with the ER-alpha negative human mammary carcinoma cell line MDA-MB-231 revealed a stimulatory effect of porcine relaxin on cellular motility.[108] In contrast, MDA-MB-231 stable transfectants secreting bioactive human proRLN2-relaxin and expressing transcripts encoding for full-size *LGR7* reveal decreased cellular motility which coincided with a downregulation of S100A4, a new relaxin target molecule belonging to the EF hand family of small calcium-binding proteins (Fig. 5).[108]

Downregulation of S100A4, also named metastasin and known to promote cellular migration,[123,124] in MDA-MB-231-RLN2 transfectants is an estrogen receptor-alpha (ERa)-independent effect of relaxin. Human breast cancer tissues display increased expression of S100A4 and metastasin is considered a marker indicative of poor clinical outcome.[125-127] S100A4 calcium-binding protein enhances p53-mediated induction of apoptosis by functionally interacting with and stabilizing the tumor suppressor protein p53.[126,128] The reported relaxin-induced transcriptional downregulation of S100A4 in MDA-MB-231 may counteract the pro-apoptotic functions of the S100A4 protein and enhance the survival of MDA-MB-231-RLN2 transfectants. Finally, RLN2 may potentially have a protective function by attenuating tissue invasiveness and metastasis of ERa-negative breast carcinoma cell populations (Fig. 6). Investigations are currently ongoing to delineate the molecular mechanisms and cytoskeletal alterations by which relaxin decreases cell motility.

Future Directions

Relaxin-like peptides have established themselves as novel multifactorial endocrine effector peptides in tumor biology. Relaxin can promote growth, differentiation and invasiveness of tumor cells in a carcinoma cell-specific context. Many details of the molecular mechanisms engaged by relaxin and INSL3 from the activation of signaling pathways to specific tumor cell function in different tumor entities are currently unknown. In particular, the signaling crosstalk in neoplastic cells between the relaxin-like ligand-receptor systems and other relevant signaling pathways, including nuclear steroid receptors and membrane-anchored tyrosine kinase receptors, is expected

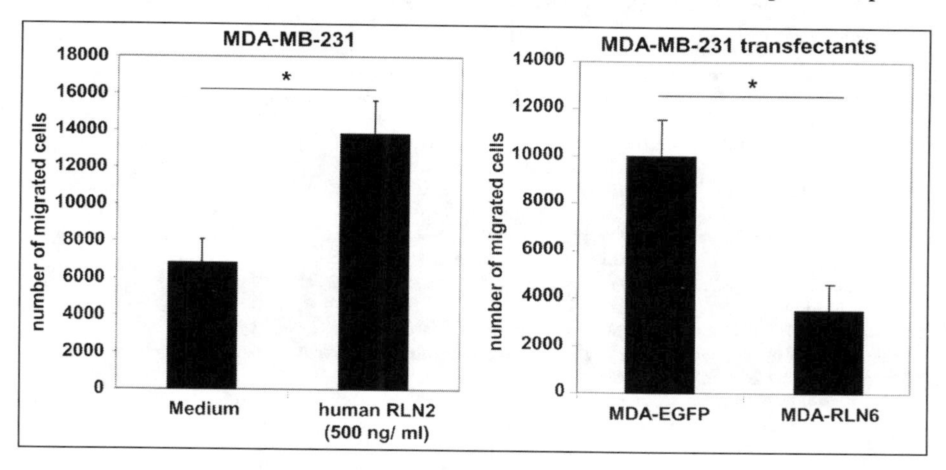

Figure 6. Short-term treatment (1-2 days) of the human ER-alpha-negative mammary carcinoma cell line MDA-MB-231 with human recombinant RLN2* resulted in enhanced motility of these breast cancer cells. By contrast, over-expression of bioactive RLN2 in stable MDA-MB-231 transfectants caused a significant decrease in motility of these carcinoma transfectants which coincided with a strong down-regulation of metastasin (S100A4) at the transcriptional and protein level. Results shown for the MDA-EGFP and MDA-RLN2-clone 6 (MDA-RLN6) clones are a representative example for all clones tested (*Human recombinant RLN2 was kindly provided by Dr. L. J. Parry, Dept. of Zoology, University of Melbourne, Australia).

to contribute substantially to our understanding of relaxin-like actions in tumor biology and beyond. Finally, recent advances in relaxin-like receptor research and newly designed synthetic relaxin-like analogues will provide the basis for the potential clinical application in the treatment of neoplastic diseases.

Acknowledgements

The authors are grateful to Christine Froehlich, Andrea Fristensky, Katrin Hammje, Elisabeth Schlüter and Dr. Ekkehard Weber for their excellent support and advise. This work has been funded by the Deutsche Forschungsgemeinschaft (DFG; KL 1249/5-1/2; HO2319/3-1) and Establishment Grants to SHK and TK awarded by the Manitoba Health Research Council (MHRC).

References

1. Tashima LS, Mazoujian G, Bryant-Greenwood GD. Human relaxins in normal, benign and neoplastic breast tissue. J Mol Endocrinol 1994; 12:351-364.
2. Mazoujian G, Bryant-Greenwood GD. Lancet 1990; 335:298-299.
3. Bongers-Binder S, Burgardt A, Seeger H et al. Distribution of immunoreactive relaxin in the genital tract and in the mammary gland of nonpregnant women. Clin Exp Obstet Gynecol 1991; 18:161-164.
4. Hombach-Klonisch S, Buchmann J, Sarun S et al. Relaxin-like factor (RLF) is differentially expressed in the normal and neoplastic human mammary gland. Cancer 2000; 89:2161-2168.
5. Alfonso P, Nunez A, Madoz-Gurpide J et al. Proteomic expression analysis of colorectal cancer by two-dimensional differential gel electrophoresis. Proteomics 2005; 5:2601-2611.
6. Stemmermann GN, Mesiona W, Greenwood FC. Immunocytochemical identification of a relaxin-like protein in gastrointestinal epithelium and carcinoma: a preliminary report. J Endocrinol 1994; 140:321-325.
7. Hombach-Klonisch S, Hoang-Vu C, Kehlen A et al. INSL3 is expressed in human hyperplastic and neoplastic thyrocytes. Int J Oncol 2003; 22:993-1001.
8. Ivell R, Hunt N, Khan-Dawood F et al. Expression of the human relaxin gene in the corpus luteum of the menstrual cycle and in the prostate. Mol Cell Endocrinol 1989; 66:251-255.
9. Klonisch T, Hombach-Klonisch S, Buchmann J et al. Relaxin-like factor (RLF) is expressed in human ovarian Sertoli-Leydig cell tumors. Fertil Steril 1999; 72:546-548.
10. Klonisch T, Ivell R, Balvers M et al. Expression of relaxin-like factor is down-regulated in human testicular Leydig cell neoplasia. Mol Hum Reprod 1999; 5:104-108.
11. Bartsch O, Bartlick B, Ivell R. Relaxin signalling links tyrosine phosphorylation to phosphodiesterase and adenylyl cyclase activity. Mol Hum Reprod 2001; 7:799-809.
12. Nguyen BT, Yang L, Sanborn BM et al. Phosphoinositide 3-kinase activity is required for biphasic stimulation of cyclic adenosine 3', 5'-monophosphate by relaxin. Mol Endocrinol 2003; 17:1075-1084.
13. Nguyen BT, Dessauer CW. Relaxin Stimulates Protein Kinase C {zeta} Translocation: Requirement for Cyclic Adenosine 3', 5'-Monophosphate Production. Mol Endocrinol 2005; 19:1012-1023.
14. Parsell DA, Mak JY, Amento EP et al. Relaxin binds to and elicits a response from cells of the human monocytic cell line, THP-1. J Biol Chem 1996; 271:27936-27941.
15. Zimmermann S, Steding G, Emmen JMA et al. Targeted disruption of the Insl-3 gene causes bilateral cryptorchism. Mol Endocrinol 1999; 13:681-691.
16. Nef S, Parada LF. Cryptorchidism in mice mutant for INSL3. Nat Genet 1999; 22:295-299.
17. Gorlov IP, Kamat A, Bogatcheva NV et al. Mutations of the GREAT gene cause cryptorchidism. Hum Mol Genet 2002; 11:2309-2318.
18. Bogatcheva NV, Truong A, Feng S et al. GREAT/LGR8 is the only receptor for insulin-like 3 peptide. Mol Endocrinol 2003; 17:2639-2646.
19. Ivell R. Biology of the relaxin-like factor (RLF). Rev Reprod 1997; 2:133-138.
20. Hsu SY, Nakabayashi K, Nishi S et al. Activation of orphan receptors by the hormone relaxin. Science 2002; 295:671-674.
21. Krajnc-Franken MA, Van Disseldorp AJ, Koenders JE et al. Impaired nipple development and parturition in LGR7 knockout mice. Mol Cell Biol 2004; 24:687-696.
22. Kawamura K, Kumagai J, Sudo S et al. Paracrine regulation of mammalian oocyte maturationand male germ cell survival. Proc Natl Acad Sci USA 2004; 101:7323-7328.
23. Guan K, Nayernia K, Maier LS et al. Pluripotency of spermatogonial stem cells from adult mouse testis. Nature 2006; 440:1199-1203.
24. Pusch W, Balvers M, Ivell R. Molecular cloning and expression of the relaxin-like factor from the mouse testis. Endocrinology 1996; 137:3009-3013.

25. Balvers M, Spiess AN, Domagalski R et al. Relaxin-like factor expression as a marker of differentiation in the mouse testis and ovary. Endocrinology 1998; 139:2960-2970.
26. Hombach-Klonisch S, Schoen J, Kehlen A et al. Seasonal expression of INSL3 and Lgr8/Insl3 receptor transcripts indicates variable differentiation of Leydig cells in the roe deer testis. Biol Reprod 2004; 71:1079-1087.
27. Cook JC, Klinefelter GR, Hardisty JF et al. Rodent Leydig cell tumorigenesis: a review of the physiology, pathology, mechanisms and relevance to humans. Crit Rev Toxicol 1999; 29:169-261.
28. Fowler KA, Gill K, Kirma N et al. Overexpression of aromatase leads to development of testicular leydig cell tumors : an in vivo model for hormone-mediated testicular cancer. Am J Pathol 2000; 156:347-353.
29. Navickis RJ, Shimkin MB, Hsueh AJW. Increase in testis luteinizing hormone receptor by estrogen in mice susceptible to Leydig cell tumors. Cancer Res 1981; 41:1646-1651.
30. Düe W, Dieckmann KP, Ley V et al. Immunohistological determination of oestrogen receptor, progesterone receptor and intermediate filaments in Leydig cell tumours, Leydig cell hyperplasia and normal Leydig cells of the human testis. J Pathol 1989; 157:225-234.
31. Young M, Lephart ED, McPhaul MJ. Expression of aromatase cytochrome p450 in rat H540 Leydig cell tumor cells. J Steroid Biochem Mol Biol 1997; 63:37-44.
32. Castle WN, Richardson JR. Leydig cell tumor and metachronous Leydig cell hyperplasia: a case associated with gynecomastia and elevated urinary estrogens. J Urol 1986; 136:1307-1308.
33. Emmen JM, McLuskey A, Adham IM et al. Involvement of insulin-like factor 3 (Insl3) in diethylstilbestrol-induced cryptorchidism. Endocrinology 2000; 141:846-849.
34. Hamdy FC. Prognostic and predictive factors in prostate cancer. Cancer Treat Rev 2001; 27:143-151.
35. Klonisch T, Müller-Huesmann H, Riedel M et al. INSL3 in the benign hyperplastic and neoplastic human prostate gland. Int J Oncol 2005; 27:307-315.
36. Cameron DF, Corton GL, Larkin LH. Relaxin-like antigenicity in the armadillo prostate gland. Ann NY Acad Sci 1982; 380:231-240.
37. Samuel CS, Tian H, Zhao L et al. Realxin is a key mediator of prostate growth and male reproductive tract development. Lab Invest 2003; 83:1055-1067.
38. Kamat AA, Feng S, Bogatcheva NV et al. Genetic targeting of relaxin and insulin-like factor receptors in mice. Endocrinology 2004; 145:4712-4720.
39. Gunnersen JM, Fu P, Roche PJ et al. Expression of human relaxin genes: characterization of a novel alternatively-spliced human relaxin mRNA species. Mol Cell Endocrinol 1996; 118:85-94.
40. Thompson V. Relaxin is upregulated during prostate cancer progression to androgen independence and is repressed by androgens. 4th International Conference on Relaxin and Related Peptides, Abstract Book 2004:O-58.
41. Janßen M, Albrecht M, Möschler O et al. Cell lineage characteristics of human prostatic stromal cells cultured in vitro. Prostate 2000; 43:20-30.
42. Silvertown JD, Ng J, Sato T et al. H2 relaxin overexpression increases in vivo prostate xenograft tumor growth and angiogenesis. Int J Cancer 2006; 118:62-73.
43. Unlu A, Leake RE. The effect of EGFR-related tyrosine kinase activity inhibition on the growth and invasion mechanisms of prostate carcinoma cell lines. Int J Biol Markers 2003; 18:139-146.
44. Madarame J, Higashiyama S, Kiyota H et al. Transactivation of epidermal growth factor receptor after heparin-binding epidermal growth factor-like growth factor shedding in the migration of prostate cancer cells promoted by bombysin. Prostate 2003; 57:187-195.
45. Jarrard DF, Blitz BF, Smith RC et al. Effect of epidermal growth factor on prostate cancer cell line PC3 growth and invasion. Prostate 1994; 24:46-53.
46. Zhang XK. Vitamin A and apoptosis in prostate cancer. Endocrine-related Cancer 2002; 9:87-102.
47. Sporn MB, Roberts AB, Goodman DS. The Retinoids, 2nd, ed. Sporn MB, Roberts AB, Goodman DS eds. New York Raven Press, 1994:319-350.
48. Thompson JN, Howell J, Pitt GAJ. Vitamin A and reproduction in rats. Proc Royal Soc 1964; 159:510-535.
49. Pasquali D, Thaller C, Eichele G. Abnormal level of retinoic acid in prostate cancer tissues. J Clin Endocrinol Metab 1996; 81:2186-2191.
50. Gyftopoulos K, Perimenis P, Sotiropoulou-Bonikou G et al. Immunohistochemical detection of retinoic acid receptor-alpha in prostate carcinoma: correlation with proliferative activity and tumor grade. Int Urol Nephrol 2000; 32:263-269.
51. Richter F, Huang HF, Li MT et al. Retinoid and androgen regulation of cell growth, epidermal growth factor and retinoid acid receptors in normal and carcinoma rat prostate cells. Mol Cell Endocrinol 1999; 153:29-38.

52. Gunnersen JM, Roche PJ, Tregear GW et al. Characterization of human relaxin gene regulation in the relaxin-expressing human prostate adenocarcinoma cell line LNCaP.FGC. J Mol Endocrinol 1995; 15:153-166.
53. Garibay-Tupas JL, Okazaki KJ, Tashima LS et al. Regulation of the human relaxin genes H1 and H2 by steroid hormones. Mol Cell Endocrinol 2004; 219:115-125.
54. Bonkhoff H, Fixemer T, Hunsicker I et al. Progesterone receptor expression in human prostate cancer: correlation with tumor progression. Prostate 2001; 48:285-291.
55. Mohler LJ, Chen Y, Hamil K et al. Androgen and glucocorticoid receptors in the stroma and epithelium of prostate hyperplasia and carcinoma. Clin Cancer Res 1996; 2:889-895.
56. Vinall R, Tepper CG, Shi XB et al. The R273H p53 mutation can facilitate the androgen-independent growth of LNCaP by a mechanism that involves H2 relaxin and its cognate receptor LGR7. Oncogene 2006; 25:2082-2093.
57. Dinjens WN, van der Weiden MM, Schroeder FH et al. Frequency and characterization of p53 mutations in primary and metastatic human prostate cancer. Int J Cancer 1994; 56:630-633.
58. Dschietzig T, Bartsch C, Stangl V et al. Identification of the pregnancy hormone relaxin as glucocorticoid receptor agonist. FASEB J 2004; 18:1536-1548.
59. Dschietzig T, Bartsch C, Greinwald M et al. The pregnancy hormone relaxin binds to and activates the human glucocorticoid receptor. Ann NY Acad Sci 2005; 1041:256-271.
60. Davies L, Welch HG. Increasing incidence of thyroid cancer in the United States, 1973-2002. JAMA 2006; 295:2164-2167.
61. Monson JP. The epidemiology of endocrine tumours. Endocrine-related Cancer 2000; 7:29-36.
62. Plunkett ER, Squires BP, Richardson SJ. The effect of relaxin on thyroid weights in laboratory animals. J Endocrinol 1960; 21:241-246.
63. Plunkett ER, Squires BP, Heagy FC. Effect of relaxin on thyroid function in the rat. J Endocrinol 1963; 26:331-338.
64. Braverman LE, Ingbar SH. Effects of preparations containing relaxin on thyroid function in the female rat. Endocrinology 1963; 72:337-341.
65. Hombach-Klonisch S, Bialek J, Trojanowicz B et al. Relaxin enhances the oncogenic potential of human thyroid carcinoma cells. Am J Pathol 2006; 169: 617-32.
66. Silvertown JD, Geddes BJ, Summerlee AJS. Adenovirus-mediated expression of human prorelaxin promotes the invasive potential of canine mammary cancer cells. Endocrinology 2003; 144:1683-1691.
67. Zarreh-Hoshyari-Khah R, Bartsch O, Einspanier A et al. Bioactivity of recombinant prorelaxin from the marmoset monkey. Regul Pept 2001; 97:139-146.
68. Vu AL, Green CB, Roby KF et al. Recombinant porcine prorelaxin produced in Chinese hamster ovary cells is biologically active. Life Science 1993; 52:1055-1061.
69. Klonisch T, Mustafa T, Bialek, J et al. Human medullary thyroid carcinoma. A source and potential target for relaxin-like hormones. Ann NY Acad Sci 2005; 1041:449-461.
70. Kamat AA, Feng S, Agoulnik IU et al. The role of relaxin in endometrial cancer. Cancer Biol Ther 2006; 5:71-77.
71. Binder C, Hagemann T, Husen B et al. Relaxin enhances in vitro invasiveness of breast cancer cell lines by up-regulation of matrix-metalloproteinases. Mol Hum Reprod 2002; 8:789-796.
72. Wyatt TA, Sisson JH, Forget MA et al. Relaxin stimulates bronchial epithelial cell PKA activation, migration and ciliary beating. Exp Biol Med 2002; 227:1047-1053.
73. Unemori EN, Lewis M, Constant J et al. Relaxin induces vascular endothelial growth factor expression and angiogenesis selectively at wound sites. Wound Repair Regen 2000; 8:361-370.
74. Khasigov PZ, Podobed OV, Gracheva TS et al. Role of matrix metalloproteinases and their inhibitors in tumor invasion and metastasis. Biochemistry 2003; 68:711-717.
75. Jiang Y, Goldberg ID, Shi YE. Complex roles of tissue inhibitors of metalloproteinases in cancer. Oncogene 2002; 21:2245-2252.
76. Kraiem Z, Korem S. Matrix metalloproteinases and the thyroid. Thyroid 2000; 10:1061-1069.
77. Lennon-Dumenil AM, Bakker AH, Wolf-Bryant P et al. A closer look at proteolysis and MHC-class-II-restricted antigen presentation. Curr Opin Immunol 2002; 14:15-21.
78. Lauritzen E, Moller S, Leerhoy J. Leucocyte migration inhibition in vitro with inhibitors of aspartic and sulphhydryl proteinases. Acta Pathol Microbiol Immunol Scand [C] 1984; 92:107-112.
79. Cataldo AM, Nixon RA. Enzymatically active lysosomal proteases are associated with amyloid deposits in Alzheimer brain. Proc Natl Acad Sci USA 1990; 87:3861-3865.
80. Adamec E, Mohan PS, Cataldo AM et al. Up-regulation of the lysosomal system in experimental models of neuronal injury: implications for Alzheimer's disease. Neuroscience 2000; 100:663-675.
81. Leto G, Tumminello FM, Crescimanno M et al. Cathepsin D expression levels in nongynecological solid tumors: clinical and therapeutic implications. Clin Exp Metastasis 2004; 21:91-106.

82. Leto G, Gebbia N, Rausa L et al. Cathepsin D in the malignant progression of neoplastic diseases (review). Anticancer Res 1992; 12:235-240.
83. Metaye T, Kraimps JL, Goujon JM et al. Expression, Localisation and Thyrotropin Regulation of Cathepsin D in Human Thyroid Tissues. J Clin Endocrinol Metab 1997; 82:3383-3388.
84. Metaye T, Millet C, Kraimps JL et al. Estrogen receptors and cathepsin D in human thyroid tissue. Cancer 1993; 72:1991-1996.
85. Kraimps JL, Metaye T, Millet C et al. Cathepsin D in normal and neoplastic thyroid tissues. Surgery 1995; 118:1036-1040.
86. Krueger S, Kellner U, Buehling F et al. Cathepsin L antisense oligonucleotides in a human osteosarcoma cell line: effects on the invasive phenotype. Cancer Gene Ther 2001; 8:522-528.
87. Kirschke H, Eerola R, Hopsu-Havu VK et al. Antisense RNA inhibaition of cathepsin L expression reduces tumorigenicity of malignant cells. Eur J Cancer 2000; 36:787-795.
88. Dohchin A, Suzuki JI, Seki H et al. Immunostained cathepsins B and L correlate with depth of invasion and different metastatic pathways in early stage gastric carcinoma. Cancer 2000; 89:482-487.
89. Urbich C, Heeschen C, Aicher A et al. Cathepsin L is required for endothelial progenitor cell-induced neovascularization. Nat Med 2005; 11:206-213.
90. Von Figura K, Hasilik A. Lysosomal enzymes and their receptors. Annu Rev Biochem 1986; 55:167-193.
91. Laurant-Matha V, Maruani-Herrmann S, Prebois C et al. Catalytically inactive human cathepsin D triggers fibroblast invasive growth. J Cell Biol 2005; 168:489-499.
92. Koike M, Shibata M, Ohsawa Y et al. Involvement of two different cell death pathways in retinal atrophy of cathepsin D-deficient mice. Mol Cell Neurosci 2003; 22:146-161.
93. Nakanishi H, Zhang J, Koike M et al. Involvement of nitric oxide released from microglia-macrophages in pathological changes of cathepsin D-deficient mice. J Neurosci 2001; 21:7526-7533.
94. Saftig P, Hetman M, Schmahl W et al. Mice deficient for the lysosomal proteinase cathepsin D exhibit progressive atrophy of the intestinal mucosa and profound destruction of lymphoid cells. EMBO J 1995; 14:3599-3608.
95. American Cancer Society, Cancer Facts and Figures, 2005. Available at http://www.cancer.org/downloads/STT/CAFF2005f4PWSecured.pdf.
96. Nardi E, Bigazzi M, Agrimonti F et al. In: Biology of relaxin and its role in the human, eds. Bigazzi, Greenwood, Gasparri; Excerpta Medica: Amsterdam 1983:417.
97. Kuenzi MJ, Sherwood OD. Monoclonal antibodies specific for rat relaxin. VII. Passive immunization with monoclonal antibodies throughout the second half of pregnancy prevents development of normal mammary nipple morphology and function in rats. Endocrinology 1992; 131:1841-1847.
98. Peaker M, Taylor E, Tashima L et al. Relaxin detected by immunocytochemistry and northern analysis in the mammary gland of the guinea pig. Endocrinology 1989; 125:693-698.
99. Hurley WL, Doane RM, O'Day-Bowman MB et al. Effect of relaxin on mammary development in ovariectomized pregnant gilts. Endocrinology 1991; 128:1285-1290.
100. Min G, Sherwood OD. Identification of specific relaxin-binding cells in the cervix, mammary glands, nipples, small intestine and skin of pregnant pigs. Biol Reprod 1996: 55:1243-1252.
101. Winn RJ, Baker MD, Merle CA et al. Individual and combined effects of relaxin, estrogen and progesterone in ovariectomized gilts. II. Effects on mammary development. Endocrinology 1994; 35:1250-1255.
102. Hwang JJ, Lee AB, Fields PA et al. Monoclonal antibodies specific for rat relaxin. V. Passive immunization with monoclonal antibodies throughout the second half of pregnancy disrupts development of the mammary apparatus and, hence, lactational performance in rats. Endocrinology 1991; 129:3034-3042.
103. Kuenzi MJ, Connolly BA, Sherwood OD. Relaxin acts directly on rat mammary nipples to stimulate their growth. Endocrinology 1995; 136:2943-2947.
104. Zhao L, Roche PJ, Gunnerson JM et al. Mice without a functional relaxin gene are unable to deliver milk to their pups. Endocrinology 1999; 140:445-453.
105. Kohsaka T, Min G, Lukas G et al. Identification of specific relaxin-binding cells in the human female. Biol Reprod 1998; 59:991-999.
106. Ivell R, Balvers M, Pohnke Y et al. Immunoexpression of the relaxin receptor LGR7 in breast and uterine tissues of humans and primates. Reprod Biol Endocrinol 2003; 1:114-127.
107. Steinetz BG, Sherwood OD, Lasano S et al. Immuno-neutralization of circulating relaxin does not alter the breast cancer-protective action of parity in MNU-treated rats. J Exp Ther Oncol 2004; 4:59-68.
108. Radestock Y, Hoang-Vu C, Hombach-Klonisch S. Relaxin downregulates the calcium binding protein S100A4 in MDA-MB-231 human breast cancer cells. Ann N Y Acad Sci 2005; 1041:462-469.
109. Bigazzi M, Brandi ML, Bani G et al. Relaxin influences the growth of MCF-7 breast cancer cells. Mitogenic and antimitogenic action depends on peptide concentration. Cancer 1992; 70:639-643.

110. Sacchi TB, Bani D, Brandi ML et al. Relaxin influences growth, differentiation and cell-cell adhesion of human breast-cancer cells in culture. Int J Cancer 1994; 57:129-134.
111. Bani D, Riva A, Bigazzi M et al. Differentiation of breast cancer cells in vitro is promoted by the concurrent influence of myoepithelial cells and relaxin. Br J Cancer 1994; 70:900-904.
112. Bani D. Relaxin and breast cancer. Bull Cancer 1997; 84:179-182.
113. Bani D, Flagiello D, Poupon MF et al. Relaxin promotes differentiation of human breast cancer cells MCF-7 transplanted into nude mice. Virchows Arch 1999; 435:509-519.
114. Hovey RC, Trott JF. Morphogenesis of mammary gland development. Adv Exp Med Biol 2004; 554:219-228.
115. Pillai SB, Rockwell C, Sherwood OD et al. Relaxin stimulates uterine edema via activation of estrogen receptors: blockade of its effects using ICI 182,780, a specific estrogen receptor antagonist. Endocrinology 1999; 14:2426-2429.
116. Pillai SB, Jones JM, Koos RD. Treatment of rats with 17beta-estradiol or relaxin rapidly inhibits uterine estrogen receptor beta1 and beta2 messenger ribonucleic acid levels. Biol Reprod 2000; 67:1919-1926.
117. Clayton H, Titley I, Vivanco MM. Growth and differentiation of progenitor/stem cells derived from the human mammary gland. Exp Cell Res 2004; 297:444-460.
118. Bryant-Greenwood GD. Relaxin as a new hormone. Endocr Rev 1982; 3:62-90.
119. Sherwood OD. Relaxin's physiological roles and other diverse actions. Endocr Rev 2004; 25:205-234.
120. Binder C, Simon A, Binder L et al. Elevated concentrations of serum relaxin are associated with metastatic disease in breast cancer patients. Breast Cancer Res Treat 2004; 87:157-166.
121. Balduyck M, Zerimech F, Gouyer V et al. Specific expression of matrix metalloproteinases 1, 3, 9 and 13 associated with invasiveness of breast cancer cells in vitro. Clin Exp Metastasis 2000; 18:171-178.
122. Ramos-DeSimone N, Hahn-Dantona E, Sipley J et al. Activation of matrix metalloproteinase-9 (MMP-9) via a converging plasmin/stromelysin-1 cascade enhances tumor cell invasion. J Biol Chem 1999; 274:13066-13076.
123. Lloyd BH, Platt-Higgins A, Rudland PS et al. Human S100A4 (p9Ka) induces the metastatic phenotype upon benign tumour cells. Oncogene 1998; 17:465-473.
124. Schmidt-Hansen B, Ornas D, Grigorian M et al. Extracellular S100A4(mts1) stimulates invasive growth of mouse endothelial cells and modulates MMP-13 matrix metalloproteinase activity. Oncogene 2004; 23:5487-5495.
125. Sherbet GV, Lakshmi MS. S100A4 (MTS1) calcium binding protein in cancer growth, invasion and metastasis. Anticancer Res 1998; 18:2415-2421.
126. Jenkinson SR, Barraclough R, West CR et al. S100A4 regulates cell motility and invasion in an in vitro model for breast cancer metastasis. Brit J Cancer 2004; 90:253-262.
127. Lee WY, Su WC, Lin PW et al. Expression of S100A4 and Met: potential predictors for metastasis and survival in early-stage breast cancer. Oncology 2004; 66:429-438.
128. Grigorian M, Andresen S, Tulchinsky E et al. Tumor suppressor p53 protein is a new target for the metastasis-associated Mts1/S100A4 protein: functional consequences of their interaction. J Biol Chem 2001; 276:22699-22708.

Chapter 9

Relaxin-Family Peptide and Receptor Systems in Brain: Insights from Recent Anatomical and Functional Studies

Sherie Ma and Andrew L. Gundlach*

Abstract

Relaxin was for many years considered primarily a hormone active within the reproductive tract with overwhelming evidence for its important roles in mammalian parturition. More recent research, however, has clearly indicated additional physiological and/or therapeutic roles for relaxin in the cardiovascular, renal and respiratory systems (see other Chapters); while a few studies have also described possible physiological effects of relaxin in the central nervous system, perhaps unsurprisingly associated with the regulation of osmotic homeostasis, blood pressure and neurohormone secretion during pregnancy and parturition. Research on relaxin and subsequently discovered, related peptides has also been particularly productive in the last five years, with some milestone discoveries (see elsewhere in this volume), including the long-awaited identification of the native receptors for relaxin and a related peptide, INSL3—the leucine-rich repeat-containing G-protein-coupled receptors-7 and -8 (LGR7/8); and the identification of a new relaxin family peptide, known as relaxin 3 and its type I G-protein-coupled receptor—GPCR135. Relaxin 3 was subsequently found to be highly conserved throughout evolution and to be the likely ancestral gene/peptide that gave rise to the current relaxin family of genes and peptides in mammals including higher primates. Interestingly, relaxin 3 and its receptor are found in highest abundance in brain, suggesting important central functions for relaxin 3/GPCR135 signaling. In this Chapter we will primarily review what is currently known about the central distribution of relaxin family peptides and their receptors and what has been described so far regarding their effects in the brain. Lastly, we will discuss likely future directions in this interesting, expanding field of research.

Discovery of Relaxin Family Genes and Peptides

With the advent of rat, mouse and human genome databases decades after the initial discovery of relaxin in 1926,[1] other members of the relaxin peptide family were quickly identified including relaxin-like factor or insulin-like peptide-3 (INSL3),[2] INSL4,[3] INSL5,[4,5] INSL6,[5-7] and more recently, relaxin 3 (or INSL7) ([1]see footnote and Table 1).[8]

[1]In this review, 'relaxin' will refer to the major stored and circulating form of relaxin (i.e., human relaxin 2, *RLN2* (gene); rodent relaxin 1, *Rln1* (gene)) and 'relaxin 3' will refer to the recently identified relaxin peptide (i.e., human and rodent relaxin 3, *RLN3/Rln3* (genes)).

*Corresponding Author: Andrew L. Gundlach, Howard Florey Institute, The University of Melbourne, Victoria 3010, Australia. Email: a.gundlach@hfi.unimelb.edu.au

Relaxin and Related Peptides, edited by Alexander I. Agoulnik.

Table 1. Relaxin and insulin-like peptide genes—their chromosomal location, principal sites of expression and their presence (or absence) in brain

Gene (abbreviation) and Alternate Names	Human Chromosomal Location	Principal Site of Expression in Mammals	Expression and/or Presence in Mammalian Brain
Insulin (INS)	11p15.5[120]	pancreas	yes
Insulin-like growth factor 1 (IGF1)	12q22[121]	liver	yes
Insulin-like growth factor 2 (IGF2)	11p15.5[122]	muscle, liver, kidney, adrenal gland in prenatal development	no
Relaxin 1 (human H1)*	9p24.1[26]	decidua and placenta (higher primates only)	no
Relaxin 2 (human H2)†	9p24.1[26]	ovary and prostate	yes
Relaxin 3 (RLX3) Insulin-like peptide-7 (INSL7)	19p13.3[9]	brain	yes
Insulin-like peptide-3 (INSL3) Relaxin-like factor Leydig-insulin-like peptide	19p13.2[2]	ovary and testis	unclear
Insulin-like peptide-4 (INSL4) Placentin, Early placental insulin-like factor	9p24[3]	placenta (higher primates only)	no
Insulin-like peptide-5 (INSL5) Relaxin-insulin-like factor-2	1p31[4]	colon and kidney	unclear
Insulin-like peptide-6 (INSL6) Relaxin-insulin-like factor-1	9p24[5]	testis	no

* No evidence of this gene in rodents. Only present in higher primates.
† Equivalent gene in rodent is known as relaxin 1 (Rln1).

Relaxin peptide has been isolated and purified from numerous mammals[9] (see Chapter 1) and phylogenic analysis of genomic and EST databases have identified additional relaxin-like sequences in nonmammalian vertebrates such as fish.[10] However, a ruminant or invertebrate relaxin gene has not been identified.

INSL3 was discovered in the early 1990's and its expression was identified in the porcine[11] and mouse[12] testis. The gene for this peptide (*INSL3*) was subsequently cloned from human and pig,[13] mouse[14,15] and rat[16] and in 2002 the mature peptide was extracted from bovine testis.[17] Shortly after, *INSL4* was identified in human,[3] but native INSL4 has not been isolated to date. *INSL5*[4,5] and *INSL6*[5-7] were discovered by EST database searching of rat, mouse and human cDNA libraries. Though native INSL5 and INSL6 peptides have not been isolated to date, these peptides have been generated synthetically and recombinantly.[18-20]

More recently, relaxin 3 (or INSL7) was identified in human, mouse[8] and rat[21] by genome database searching. Shortly after, native relaxin 3 peptide was isolated from pig brain.[22] In contrast to relaxin, relaxin 3-like sequences appear prior to the divergence of teleosts, such as the puffer fish (*Takifugu rubripes*) and the highly conserved nature of relaxin 3 through evolution suggests the original function of relaxin 3 was likely in the brain and any role in reproductive functions was acquired just prior to the divergence of amphibians (see Chapter 1).[10,23] This work also highlights the strongly conserved and therefore likely important function of relaxin 3 as a neuropeptide and raises the important question of whether it acts independently or in conjunction with relaxin, which is also produced by small populations of neurons in mammalian brain (see later discussion).

Human and Rodent Relaxin Family Peptides—Gene Characteristics and Biosynthesis

Over a 20 year period, analysis of genomic clones encoding human relaxin,[24] genomic Southern blot analysis[25] and searches of the human genome database[8] have revealed the existence of three genes that encode human relaxins—*RLN1*, *RLN2* and *RLN3*, respectively. The peptide encoded by *RLN2* is the predominant stored and circulating form of relaxin in humans. The non-allelic *RLN1* and *RLN2* genes are clustered together on human chromosome 9 at 9p24 together with *INSL4* and *INSL6* (see Table 1),[8,26] while the more recently identified *RLN3* gene is localized on human chromosome 19 at 19p13.3, which is in close proximity to *INSL3* (19p13.2; see Table 1).[2,8] This may explain why *RLN1* and *RLN2* share approximately 90% sequence homology,[26] but there is very little homology between the *RLN2* and *RLN3* sequences, other than that encoding the essential Arg-X-X-X-Arg-X-X-Ile receptor binding motif of the B-chain and key structural elements in the A-chain.[9] This data suggests that *RLN1* and *RLN2* were derived through recent duplication events and are evolutionarily divergent to *RLN3*. Extensive searches of the genome and Celera Genomics databases have revealed that there is unlikely to be an equivalent of *RLN1* in species other than higher primates.[27] Concurrent with the identification of *RLN1* and *RLN2*, an *RLN2* homologue was identified in mouse and rat genomes and designated *Rln1*. With the discovery of *RLN3*, equivalent mouse and rat sequences were identified and, for consistency, were designated as *Rln3*.[8,21] The mouse *Rln1* and *Insl6* genes are located at a similar region of chromosome 19B, whereas *Rln3* is located on mouse chromosome 8 at 8C2, 13Mb from mouse *Insl3* at 8B3.2 (see Table 1).[8,28] Again, there is very little homology between mouse *Rln1* and *Rln3* other than the key core binding elements required of relaxins in the B-chain and key structural elements in the A-chain.[8] Somewhat surprisingly, relaxin sequences in different mammalian species vary by as much as 42-60%, despite a high similarity in biological activity.[29] Nonetheless, it is proposed that relaxin receptor signaling may be fairly well conserved among species (see below). *RLN2* and *Rln1* share 67% and 54% homology in the A- and B-chains respectively.[30] Interestingly though, the relaxin 3 gene appears to be evolutionarily conserved with 73% homology between *RLN3* and *Rln3* of the C-peptide domain and 93% homology in the A- and B-domains.[8] There is little homology however, of the C-peptide domain (<20%) of relaxin 3 and other insulin- and relaxin-family members in human and mouse, which may reflect differences in function(s) of the C-peptide.

During the 1980's it was discovered that relaxin peptides, like many hormones, are secreted as a precursor peptide of approximately 23 kDa in size.[31,32] The single chain precursor, preprorelaxin, consists of a signal peptide, B-chain, connecting peptide (C-peptide) and A-chain. Processing in luteal cells in vitro involves cleavage of the 3 kDa signal peptide from the large polypeptide to form prorelaxin.[31,32] This cleavage occurs as the emerging peptide chain is translocated across the endoplasmic reticular membrane and studies of a transformed *Escherichia coli* strain[33] indicate that the resultant prorelaxin peptide is subsequently packaged into secretory granules of the Golgi apparatus.

Very recent studies indicate that *Insl6* is primarily expressed in meiotic and postmeiotic germ cells of the testis, where it is a secreted protein localized to the endoplasmic reticulum and Golgi.[18] Processing of mouse and human INSL6 in transfected CHO cells is furin-dependent and posttranslational modifications include the presence of disulfide bonds, glycosylation and ubiquitination.[18]

Interestingly, nothing is known of the function of the large relaxin C-peptides, but it is possible that they have other activities, as this is the case for the precursor of the structurally-related insulin. At the time of writing there have been no investigations to examine the process of relaxin biosynthesis in neurons (but see below).

Relaxin Family Peptide Receptors: Discovery, Distribution and Structure-Activity

The identity of the first relaxin receptor was discovered in 2002. Two orphan receptors, designated leucine-rich repeat-containing G-protein coupled receptor-7 (LGR7) and -8 (LGR8), were found to be responsive to relaxin.[34] Thus, somewhat contrary to predictions of many investigators in the field[35] and *some* suggestive experimental data,[36,37] the receptor for relaxin was not a membrane-associated tyrosine kinase like those that bind the structurally similar insulin and IGF-1. LGR7 and LGR8 are structurally related to the GPCR superfamily of glycoprotein hormones such as luteinizing hormone, follicle-stimulating hormone and thyrotropin-stimulating hormone.[34,38] They are seven transmembrane-spanning GPCRs and have a large N-terminal extracellular ectodomain consisting of leucine-rich repeats, which are postulated to form a horseshoe-shaped interaction motif for ligand binding observed in other LGRs of this family. They also contain a unique low-density lipoprotein receptor-like cysteine-rich (LDL) motif at the N-terminus. Porcine relaxin applied to LGR7- and LGR8-transfected cells results in dose-dependent cAMP generation, with median effective concentrations of 1.5 and 5.0 nM respectively.[34] Human LGR7 binds human or porcine relaxin with high affinity (EC_{50} < 1 nM)[34,39] and human relaxin 3 with lower affinity (EC_{50} ~ 20 nM),[39] whereas human LGR8 binds relaxin (EC_{50} ~10 nM), but not relaxin 3.[34,39]

Insulin like-peptide 3 (INSL3) was subsequently reported to specifically interact with LGR8 in vitro, suggesting that INSL3 is the preferred native ligand for LGR8.[40,41] INSL3 activation of LGR8 has been shown to initially involve $G\alpha_s$ coupling followed by an inhibitory $G\alpha_{OB}$ pathway.[42] This signaling system is important in testis descent[43] and mammalian oocyte maturation and male germ cell survival.[44,45] In fact, deletion of *Insl3*[46,47] or *Lgr8*[47,48] genes results in the same phenotype—cryptorchidism—providing further convincing evidence for the native INSL3-LGR8 pairing. Importantly with regard to the focus of this chapter, in contrast to relaxin (see below), there is no clear evidence to date for INSL3 expression by neurons in the mammalian brain despite good evidence for the presence of LGR8 in brain;[49-51] so further studies are required of this issue.

Shortly after the identification of LGR-7 and -8 as relaxin-peptide receptors, two additional orphan G-protein-coupled receptors (GPCRs)—GPCR135 and GPCR142—were found to selectively respond to recombinant human relaxin 3.[22,52] GPCR135, also known as somatostatin- and angiotensin-like peptide receptor (SALPR[53]), is a type-I GPCR with significant homology to the somatostatin receptor (SSTR5) and angiotensin II receptor (AT1) with 35% and 31% identity, respectively.[22,53] GPCR135 gene sequences have been identified in the human, rat and mouse genomes.[22] This receptor binds human relaxin 3 with high affinity in vitro (K_d = 300 pM) and functional activation results in inhibition of cAMP generation in transfected Chinese hamster ovarian cells (CHO-K1; EC_{50} 200-300 pM), suggesting coupling of GPCR135 to $G_{i/o}$.[22] Importantly, GPCR135 does not bind human relaxin or any other member of the insulin/relaxin peptide family tested so far.[22]

The related receptor GPCR142 bound [^{125}I]-labeled and unlabeled human relaxin 3 ($K_d \approx$ 2 nM), but was initially reported as unresponsive to some other members of the insulin/relaxin family.[52] Subsequent molecular studies have demonstrated that INSL5 is the preferred ligand for GPCR142.[19,54] Consistent with this, relative levels of INSL5 and GPCR142 expression are both highest in the gastrointestinal tract in humans. In relation to possible roles in the brain, neither INSL5 or GPCR142 mRNA is detectable in abundance in mouse brain[19,54] and although human and mouse genes for GPCR142 have been isolated, only a pseudogene is present in the rat genome.[52] In mouse brain sections, a hydrophilic relaxin 3-INSL5 chimeric peptide ([^{125}I]-R3/I5) consisting of the A-chain of INSL5 and B-chain of relaxin 3 showed selectivity for GPCR135 over LGR7,[55] whereas no GPCR142-specific [^{125}I]-INSL5 binding sites are detectable, offering further support for the importance of the relaxin 3/GPCR135 ligand-receptor pairing in rat, mouse and human brain. Lastly, the identity of the receptors for INSL4 and INSL6 is currently unknown and little is known about the possible presence, distribution and role of these peptides in the brain. Current knowledge of the various peptide-receptor relationships is summarized elsewhere in this book.

Relaxin and Related Peptides: Peripheral and Central Distribution

The richest source of relaxin in mammals is the corpus luteum of the pregnant and nonpregnant female and relaxin reaches its highest concentration in plasma during pregnancy (see Chapter 2).[9,56] In the male, the prostate gland is the primary source of relaxin production, where the hormone has an important function in prostatic growth and fertility.[57,58] This is consistent with detection of LGR7 transcripts in ovary, uterus, testis and prostate gland[34] and suggests roles for relaxin-LGR7 in the female and male reproductive tract. LGR7 mRNA is also detected in human kidney, adrenal gland, heart and brain (the latter two of which also express relaxin), suggesting further roles for relaxin in the renal-cardiovascular and central nervous systems.

In contrast, relaxin 3 mRNA expression is highest in rat, mouse and human brain,[8,21,22] consistent with a 'neuropeptide' role for relaxin 3. Lower levels of relaxin 3 mRNA have been detected in mRNA extracted from mouse spleen, thymus, lung, testis and ovary[8] and in human testis, but only using a more sensitive RT-PCR assay.[22]

The critical role of INSL3 in testicular descent has been well established and mutation of the INSL3 or LGR8 gene, results in a cryptorchidism phenotype in mice and humans[59-62] (see Chapter 3) and stimulates oocyte maturation and suppresses male germ cell apoptosis.[44] Although *Lgr8* mRNA is highly expressed in thalamic nuclei of rat brain, expression of *Insl3* mRNA in the central nervous system is not detectable using similar methods,[49] suggesting that expression levels are very low or that the source of central INSL3 is peripheral tissues such as the testis and ovary. Further research is required to clarify this issue.

INSL5 and its receptor GPCR142 are highly expressed in the gastrointestinal tract, whereas *Insl5* mRNA is not readily detectable in adult mouse brain by Northern blot analysis, though high levels of *Insl5* mRNA expression have been observed in fetal brain and pituitary with barely detectable levels of GPCR142 mRNA expression.[19,54] The *Insl5* and *Gpcr142* genes are absent from and a pseudogene in, the rat genome, respectively.[10,52,54] However, a recent study did report the presence of *Insl5* mRNA in mouse hypothalamus using RT-PCR and the presence of INSL5-like immunoreactivity in magnocellular neurons of the paraventricular, supraoptic, accessory secretory and supraoptic retrochiasmatic nuclei and in nerve processes in the median eminence.[20] At present though, these immunohistochemical results await independent confirmation using in situ hybridization histochemistry that these neurons express *INSL5* mRNA or studies to reveal that they selectively accumulate the peptide and further control tests to rule out any possible cross-reactivity of the polyclonal antisera to other antigens. Details on the presence of INSL4 and INSL6 in brain are not available at present. For these reasons, the remainder of this section will focus on central relaxin and relaxin 3 systems.

In the rat brain, relaxin mRNA is expressed in forebrain regions including the anterior olfactory nucleus, lateral orbital cortex, tenia tecta, piriform cortex, neocortex and dentate gyrus and CA1-3 fields of the hippocampus (Fig. 1).[63-65] A similar expression pattern for relaxin mRNA was reported in preliminary studies of the mouse brain[64,66] but to date there have been no detailed studies published on the mouse or the distribution in human brain. In the developing rat, expression of relaxin mRNA was undetectable at embryonic day 15, but detectable at postnatal day 1 in the same regions observed in adult male and female brain.[63]

Studies in our laboratory using a polyclonal antibody raised against mature relaxin peptide revealed 'relaxin-like' immunoreactivity (-LI) in cell bodies of the regions listed above[65] (Fig. 1), but relaxin-LI was not readily detected in nerve fibers or terminals, indicating the presence of low relaxin levels in nerve terminals that were undetectable by current methods, or reflecting the existence of a different process for relaxin utilization in neurons to that for other neuropeptides. For example, relaxin normally stored in neuronal soma might require a particular type or strength of stimulus to initiate axonal transport and remote release, as suggested in a recent review on neuropeptide actions.[67] Alternatively, relaxin, like hormones such as arginine-vasopressin and oxytocin, may produce effects in the brain more by somatodendritic release and diffusion or 'volume transmission'.[68] Interestingly, the relative densities of relaxin mRNA and relaxin-LI differed in various forebrain regions. Neurons in the outer layers of the piriform (olfactory) cortex appeared

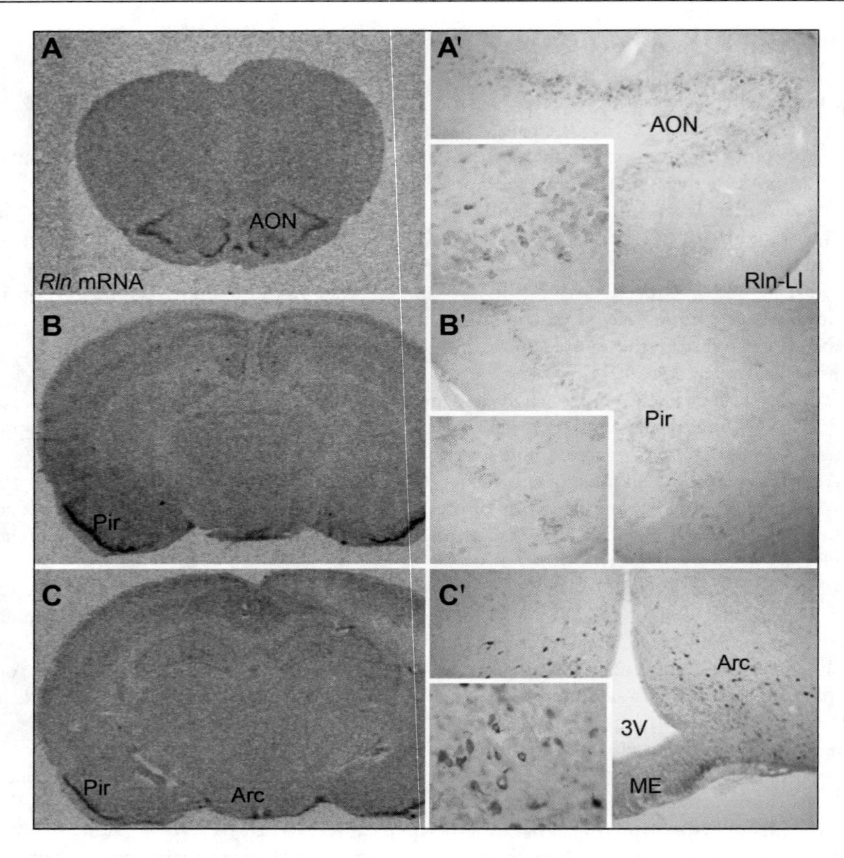

Figure 1. Distribution of relaxin (*Rln*) mRNA (A-C) and Rln-like immunoreactivity (-LI) (A′-C′) in different regions of the rat forebrain. Autoradiograms of coronal sections of rat brain illustrate the regional localization of relaxin mRNA in (A) anterior olfactory nucleus (AON); (B) piriform cortex (Pir); (C) arcuate nucleus (Arc); and the corresponding photomicrographs illustrate the distribution of Rln-LI in these areas at low- (A′-C′) and high-power (insets) magnification. Nonspecific hybridization in the presence of a 100-fold excess of unlabeled oligonucleotides was equivalent to tissue background and no immunostaining was observed in the absence of primary antibody (data not shown). Modified from Figure 2 of Ma et al[65] with permission; ©2005 Blackwell Publishing.

to express the highest levels of relaxin mRNA, whereas levels of relaxin-LI in this region were relatively low; and the reverse was observed in the hypothalamic arcuate nucleus, with moderate levels of relaxin mRNA and prominent levels of relaxin-LI. This may indicate regional differences in relaxin utilization—i.e., local production and stimulus-dependent local or remote release in piriform neurons and local production and high ongoing, adjacent or distant release in arcuate cells. An earlier study using the same antisera (AS#2) also reported relaxin-LI only in scattered neurons of the arcuate nucleus with no relaxin-containing nerve fibers or terminals detected.[57] In this case, staining was observed in the soma of neurons, regardless of whether the antiserum was raised against the mature peptide (AS#2) or against the C-peptide fragment (AS#325).[57] Furthermore, the level of staining and the number of stained cells in the arcuate were not significantly altered in pregnant, late-pregnant and nonpregnant rats,[57] suggesting that the function of central relaxin in this hypothalamic region may not be directly involved in the maintenance of pregnancy or that modulation of relaxin-LGR7 signaling occurs at other points in the hypothalamic-pituitary-gonadal (HPG) axis and during pregnancy.

In rat and mouse brain, relaxin 3 is abundantly expressed in the pontine region known as the *nucleus incertus* (NI; Fig. 2).[21,69,70] A similar delineation of this structure was reported in the cat,[71] though it is also known as *locus incertus* in the golden hamster,[72] *nucleus O* and central gray pars alpha in rabbit[73] and rat,[74] *nucleus recessus sulci mediani* in the rat[75] and the ventromedial part of the dorsal tegmental in rat and rabbit.[76] Recent interest in the NI stems from its abundant expression of type-1 corticotropin-releasing hormone receptor (CRH-R1)[77] and due to its anatomical location adjacent to the fourth ventricle, it is postulated to be a site at which CRH circulating in the CSF may elicit additional central effects related to stress. In this regard, recent studies have revealed that relaxin 3- and CRH-R1-IR are colocalized in NI neurons,[69] strongly suggesting that CRH can directly alter relaxin 3 neuron activity and that this can modulate animal behavior in specific ways linked to the stress response (see below).

Interestingly, the efferent and afferent connections of the NI strongly implicate it as a key modulatory/relay center able to influence processes related to *"behavioral planning, habenular function, hippocampal and cortical activity involved in attention and memory and oculomotor control"*.[78] Furthermore, the NI may be involved in two interconnected circuits involving its projections to the interpeduncular nucleus, habenula, mammillary bodies, suprachiasmatic nucleus, septum and hippocampus that may modulate circadian and theta rhythms.[79]

Immunohistochemical mapping experiments in our laboratory have used a polyclonal antibody raised against a region of the relaxin 3 C-peptide that is highly homologous in human, rat and mouse; and are therefore indicative of the prepro- or pro-form of the peptide.[70] These studies revealed a map of relaxin 3-containing axon and fiber projections (Fig. 2) that very closely matches that of projections of the NI determined using anterograde tract-tracing methods[78,79] and a similar mapping using a monoclonal antibody against the N-terminal of the A-chain of mature relaxin 3.[69] Thus, cell bodies of the NI contained dense immunostaining and relaxin 3-LI was detected in nerve processes throughout mid- and fore-brain areas such as hypothalamus, thalamus and limbic and cortical structures. Along with CRH-R1,[69] double-label immunofluorescence revealed that most, if not all, relaxin 3 neurons also express glutamic acid decarboxylase (GAD65; Fig. 2).[70] Tanaka et al[69] also demonstrated using electron microscopy that relaxin 3-LI was localized in dense-core vesicles in NI perikarya and synaptic terminals.

Similar staining patterns in brain with antibodies against the C-peptide and A-chain of relaxin 3 indicate that the C-peptide of relaxin 3 is axonally transported and that further processing of the peptide from precursor to mature form may occur within the dense-core vesicles during axonal transport and/or in the nerve terminal or after secretion. Currently nothing is known of the specific process of relaxin peptide biosynthesis in neurons and the metabolic fate or function of the large C-peptide, but it is possible that it has biological activity, as is the case for other similar brain peptides.

Distribution of LGR7, LGR8 and GPCR135 in Rodent Brain

Several relaxin-family peptide receptors are broadly distributed in the brain (Table 2).[80] LGR7 mRNA is present in human brain *and* in kidney, testis, placenta, uterus, ovary, adrenal, prostate, skin and heart.[34] Moreover, LGR8 transcripts are present in human brain *and* in kidney, testis, uterus, thyroid, muscle, peripheral blood cells and bone marrow.[34] Prior to the recent molecular identification of the relaxin receptors, autoradiography of [^{32}P]-labeled B29-human relaxin was employed in the first studies to visualize the distribution of relaxin binding sites in the rat brain.[81] This technique revealed specific, high-affinity relaxin binding in various discrete nuclei of the olfactory system, neocortex, hypothalamus, hippocampus, thalamus, amygdala, midbrain and medulla of male and female rats. Relaxin binding sites were displaceable by 100 nM unlabeled H2 relaxin, but not 100 nM of insulin, nerve growth factor, angiotensin II, or atrial natriuretic peptide.[81] However, following the identification of LGR7 and GPCR135 cDNA sequences a more detailed anatomical mapping of relaxin receptors in brain was possible. Studies in our laboratory

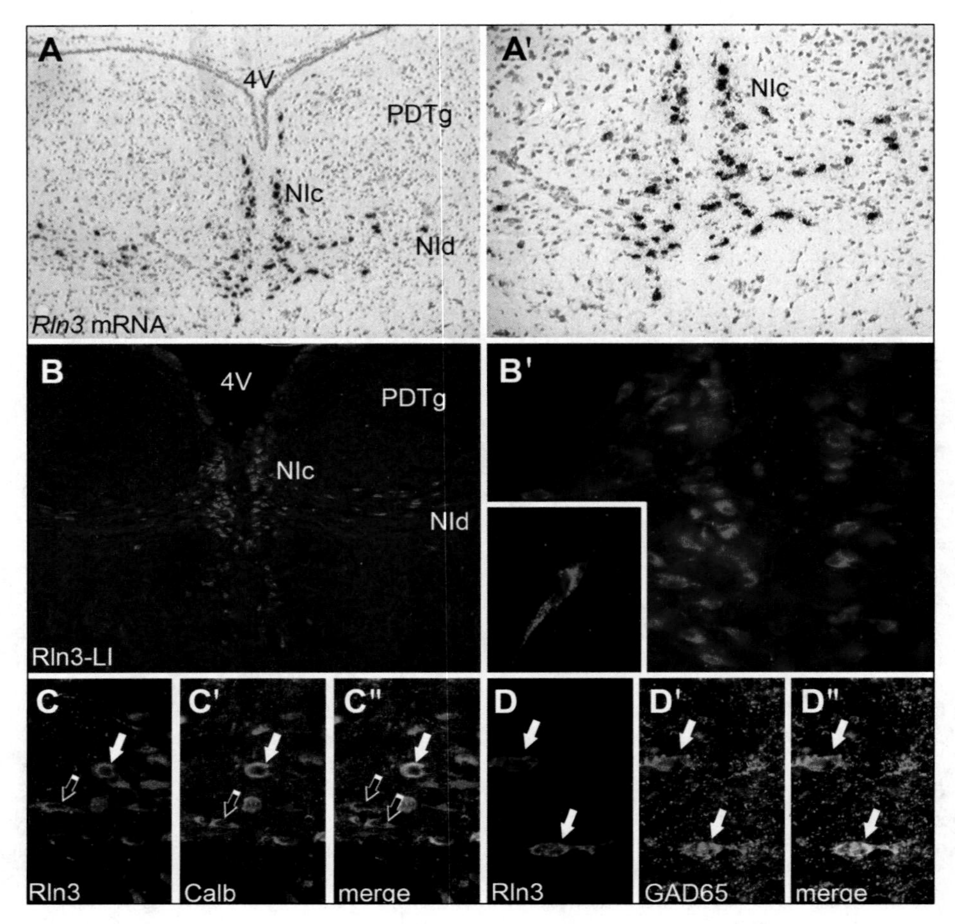

Figure 2. Relaxin 3 (*Rln3*) mRNA and double-label immunofluorescence of Rln3 with GAD and calbindin in neurons of the rat nucleus incertus (NI). Nuclear-emulsion autoradiograms confirm that *Rln3* mRNA is associated with individual neurons of the NI at low- (A) and high-power (A') magnification. Confocal photomicrographs illustrate localization of Rln3-LI in neuronal cell bodies of the NI at low- (B) and high-power (B') magnification. Double-label immunofluorescence revealed colocalization of Rln3-LI with (C,C',C") calbindin and (D,D',D") GAD65 (white arrows). Most, if not all, Rln3-LI cells were GAD-positive. Numerous calbindin-positive only and Rln3-positive only cells were detected (open arrows).

revealed that the distribution of [^{33}P]-labeled human relaxin binding in rat brain closely corresponds with the distribution of LGR7 mRNA (Fig. 3),[82] whereas the distribution of specific binding sites for a radiolabeled, hydrophilic relaxin 3 chimeric analogue ([^{125}I]-R3/I5) correlates with that of GPCR135 mRNA (Fig. 3).[70,83] This data provides further evidence for the native relaxin-LGR7 and relaxin 3-GPCR135 pairings.

To further identify the localization and role of LGR7 in mice, a strain of LGR7-knockout mice was created with a *LacZ*/MC1-neocassette inserted in place of the deleted sequence, such that the *LacZ* reporter gene was in frame with the LGR7 coding sequence.[84] This allowed the simple analysis of more than 40 different tissues isolated from male and female mutant mice for β-galactosidase (β-Gal) activity, using standard staining methods. These studies revealed β-Gal

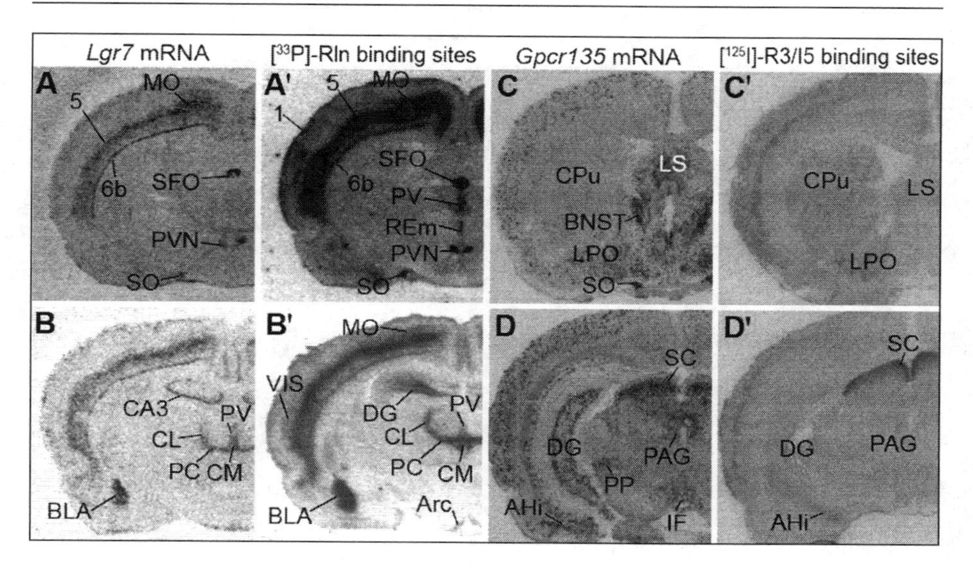

Figure 3. Comparative distribution of (A,B) *Lgr7* mRNA and (A′,B′) corresponding [^{33}P]-Rln binding sites; and (C,D) *Gpcr135* mRNA and (C′,D′) corresponding [^{125}I]-R3/I5 binding sites in rat forebrain. Strong levels of *Lgr7* mRNA were detected in magnocellular neurons in the supraoptic nucleus (SON) and paraventricular nucleus (PVN), layers 5 and 6b of neocortex and motor (MO) and visual (VIS) cortices, the subfornical organ (SFO), the basolateral amygdala (BLA), centromedial- (CM), paracentral- (PCL), centrolateral- (CL) and paraventricular- (PV) thalamic nuclei, the hippocampus (CA3) and arcuate nucleus (ARC). Most of these brain regions also contain high densities of [^{33}P]-RLN binding sites. Levels of nonspecific hybridization and nonspecific binding were both equivalent to film background (data not shown). *Gpcr135* mRNA was associated with scattered neurons of the cortex, lateral septum (LS), bed nucleus of the stria terminalis (BNST), lateral preoptic (LPO) and supraoptic (SO) hypothalamic nuclei, superior colliculus (SC), dentate gyrus (DG), periaqueductal grey (PAG), peripeduncular nucleus (PP), infrafascicular nucleus (IF) and the amygdalohippocampal transition area (AHi). Most of these regions also contained [^{125}I]-R3/I5 binding sites though some regions show poor correlation. Modified from Figure 2 of Ma et al[82] and Figure 9 of Ma et al[70] with permission; ©2006 and 2007 Elsevier Ltd.

staining in uterus, heart, oviduct, mammary gland, testis and brain.[84,85] Interestingly, high levels of β-Gal staining were observed in the anterior pituitary gland of pregnant mice, but not in non-pregnant and male mice, suggesting a function for relaxin-LGR7 systems in hormonal release from the pituitary during pregnancy.

The expression of LGR8 mRNA in rat brain is relatively enriched in the posterior and intralaminar thalamic nuclei.[49] Though an identified source of INSL3 in the brain is yet to be found, thalamic neurons that express LGR8 appear to respond to INSL3 in rat brain slice preparations.[49] Recent studies in our laboratory using [^{125}I]-labeled-INSL3 suggest that in addition to LGR8 on the soma of thalamic neurons, these receptors are also present on ascending axons and terminals in basal ganglia.[50,51] Therefore it is postulated the INSL3/LGR8 signaling may be related to sensorimotor control in the rat and other species.

Using RT-PCR, *GPCR135* mRNA was detected in human brain and in adrenal gland, testis and thymus, with corresponding relaxin 3 mRNA expression found in human brain and testis.[22] In the mouse, relaxin 3 mRNA is present at highest levels in the brain, with lower expression reported in spleen, thymus, lung and ovary.[8] In situ hybridization histochemistry of *Gpcr135* mRNA in rat brain revealed abundant expression in the olfactory bulb, hypothalamus (paraventricular and supraoptic nuclei, preoptic and posterior areas), hippocampus, septum and extended

Table 2. The relaxin family peptide receptors[80]

Receptor (IUPHAR Name; Alternate Name)	Agonist Peptide (Rank of Potency*)[80]	Expression in Mammalian Brain
LGR7 (RXFP1)	H2>H1>H3>>INSL3	yes[82]
LGR8 (RXFP2)	INSL3>H2=H1>>H3	yes[50,51]
GPCR135 (RXFP3; SALPR)	H3 (insensitive to all other relaxin-family ligands)	yes[70,83]
GPCR142 (RXFP4; GPR100)	INSL5=H3	no[54]

*Agonist potency based on in vitro receptor signaling of human cloned receptor. H1, human relaxin-1; H2, human relaxin-2; H3, human relaxin 3

amygdala, with lower levels present in the cerebral cortex, lateral periaqueductal gray, nucleus incertus and central gray regions of the brainstem (Fig. 3).[22,83] Autoradiography of [^{125}I]-relaxin 3 binding to tissue sections proved to be difficult as the ligand generated high nonspecific binding and interpretation of such results may have been difficult also as the ligand was theoretically capable of binding to both GPCR135 and LGR7. Liu and colleagues[55,83] overcame this problem by creating a chimeric peptide with the A-chain of INSL5 and B-chain of relaxin 3 (R3/I5) that selectively bound GPCR135 and GPCR142 compared to LGR7. Autoradiographic detection of [^{125}I]-R3/I5 binding revealed a distribution that closely correlated with that of GPCR135 mRNA in rat brain (Fig. 3).[70,83] High densities of both mRNA and binding sites were observed in the olfactory bulb and anterior olfactory nucleus. Low to moderate levels of binding sites and mRNA were detected in the lateral septum. Moderate-high levels of both mRNA and binding sites were reported in the paraventricular and supraoptic hypothalamic nuclei with lower levels in the lateral hypothalamus. Moderate-high levels of mRNA and binding sites were observed in various thalamic nuclei, within the hippocampal formation, habenula, amygdala, superior colliculus, interpeduncular nucleus, dorsal raphé and periaqueductal gray. In addition, moderate densities of binding sites were observed in motor, somatosensory, temporal and visual cortices.[70,83] In the pons, both mRNA and binding sites were detected in the nucleus incertus, spinal trigeminal tract and nucleus of the solitary tract.[70,83]

Brain regions in the rat that display both GPCR135 and LGR7 include the olfactory bulb, anterior olfactory nuclei, paraventricular and supraoptic hypothalamic nuclei, certain midline thalamic nuclei, hippocampus and dentate gyrus and interpeduncular nucleus,[70,82,83] so it is possible that the relaxin and relaxin 3 peptide-receptor systems may interact in concert to modulate the function of neurons in these regions. The available anatomical evidence does suggest that any interaction between GPCR135 and LGR7 signaling may be indirect however, as in many regions such as the cortex, amygdala and hippocampus, it is likely that LGR7 and GPCR135 are expressed on different populations of cells—i.e., the pattern of receptor distribution suggests LGR7 is present on glutamatergic pyramidal neurons and GPCR135 is present on GABAergic interneurons. Whether relaxin *and* relaxin 3 both interact with LGR7 in vivo may depend on the local concentration of peptide, as human relaxin 3 preferentially activates human GPCR135 *cf.* LGR7 (EC_{50} <1 nM *vs* ~ 20 nM),[22,39] but in support of this possibility, it does appear that relaxin 3 is far more abundant in brain than relaxin.

Central Actions of Relaxin and Relaxin 3

Hemodynamic Effects

Relaxin has a well characterized dipsogenic effect in the brain following iv[86] or icv administration.[87] LGR7 is present in regions with known functions in central hemodynamic control, including hypothalamic and basal forebrain nuclei such as the organum vasculosum of the lamina terminalis, subfornical organ and the supraoptic and paraventricular nuclei.[81,82] The former two structures are circumventricular organs that lie along the rostral border of the third ventricle, effectively outside blood brain barrier and have direct exposure to blood-borne peptides and hormones in the cerebral spinal fluid (due to fenestrated endothelial cells of the blood vessels of this region).[88] These brain nuclei also have axonal projections to other regions that subserve vasopressin secretion and water drinking (such as the median preoptic, supraoptic and paraventricular nuclei) and relaxin is thought to activate these brain pathways via angiotensin II signaling.[86,89] Relaxin 3 has also been demonstrated to induce drinking in rats when administered icv, though with lower potency to relaxin.[90] This action may be due to its lower affinity for LGR7,[90] although it is currently unclear if this effect is partially mediated by GPCR135 signaling within similar hypothalamic and other nuclei, including the subfornical organ.[66]

In the pregnant rat, injection into the brain of a specific relaxin monoclonal antibody negated the normal increase in drinking that was observed during the second half of pregnancy.[87] Instead, there was a significant decrease in water consumption at night, but not during the day.[87] The dipsogenic effect of exogenous relaxin in the rat also varies with time of injection during the light-dark cycle, where dose-dependent water drinking caused by relaxin is maximal at night compared to consumption during the day. This effect while perhaps predictable in nocturnal animals such as rats, may nonetheless suggest an involvement of relaxin in this and other circadian-based functions[87,91] (see below).

Neuroendocrine Effects

Neuroendocrine actions of relaxin in the rat are presumed to be primarily mediated via changes in oxytocin and vasopressin secretion from hypothalamic-hypophyseal neurons. Relaxin infusion (icv) in anaesthetized, lactating rats results in significant inhibition of suckling-induced reflex milk-ejection and this effect is reversed by naloxone, suggesting an involvement of endogenous opioid systems,[92] although the sites of action and precise downstream involvement of opioid-receptor signaling are unclear.

Acute systemic administration of relaxin causes significant increases in systolic and diastolic blood pressure in pregnant and lactating rats, together with elevation of plasma vasopressin and oxytocin concentrations.[93] Central and systemic administration of relaxin significantly elevates Fos-like immunoreactivity in oxytocin cells in the supraoptic nucleus, magnocellular cells of the rostral and caudal hypothalamic paraventricular nucleus and dorsal part of the organum vasculosum of the lamina terminalis,[94,95] suggesting that these neurons are directly (or indirectly) activated by both central and circulating relaxin to mediate these hypothalamic effects.

An early report by Cronin and Malaska[96] demonstrated that in anterior pituitary cells from adult female rats, relaxin enhanced levels of cAMP generation and this effect was amplified in the presence of pertussis toxin, indicating the involvement of an active G-protein in the signaling. In addition, anterior pituitary cells from adult males exhibited a mean two-fold maximal stimulation after relaxin, compared with a six-fold increase measured in cells from female rats. Relaxin stimulation in these cells was significantly inhibited by dopamine and somatostatin, two key hypophysiotrophic factors.[96] Thus, relaxin is a potent, specific and reversible stimulator of G-protein coupled cAMP production in cultured anterior pituitary cells and the effect is enhanced in females and may involve hypothalamic hormones.

Recent studies have revealed strong effects of relaxin 3 on feeding in rats (Fig. 4). Infusions of commercially-produced, synthetic relaxin 3 (180 pM; icv) in satiated rats increased 1-h food intake post administration in the early light phase (~ 90%) and the early dark phase (~ 50%).[97] In line with the presence of high concentrations of GPCR135 in the region, infusions in the paraventricular nucleus (PVN) of relaxin 3 (18 pM) significantly increased 1-h food intake in satiated rats in the early light phase (~ 250%) and the early dark phase (~ 50%).[97] Repeated intra-PVN injections of relaxin 3 (180 pmol/twice a day for 7 days) significantly increased cumulative food intake by ~ 20% in *ad libitum* fed animals compared to vehicle-treated controls, with an associated increase in plasma leptin and decrease in plasma thyroid-stimulating hormone (TSH) concentrations.[98]

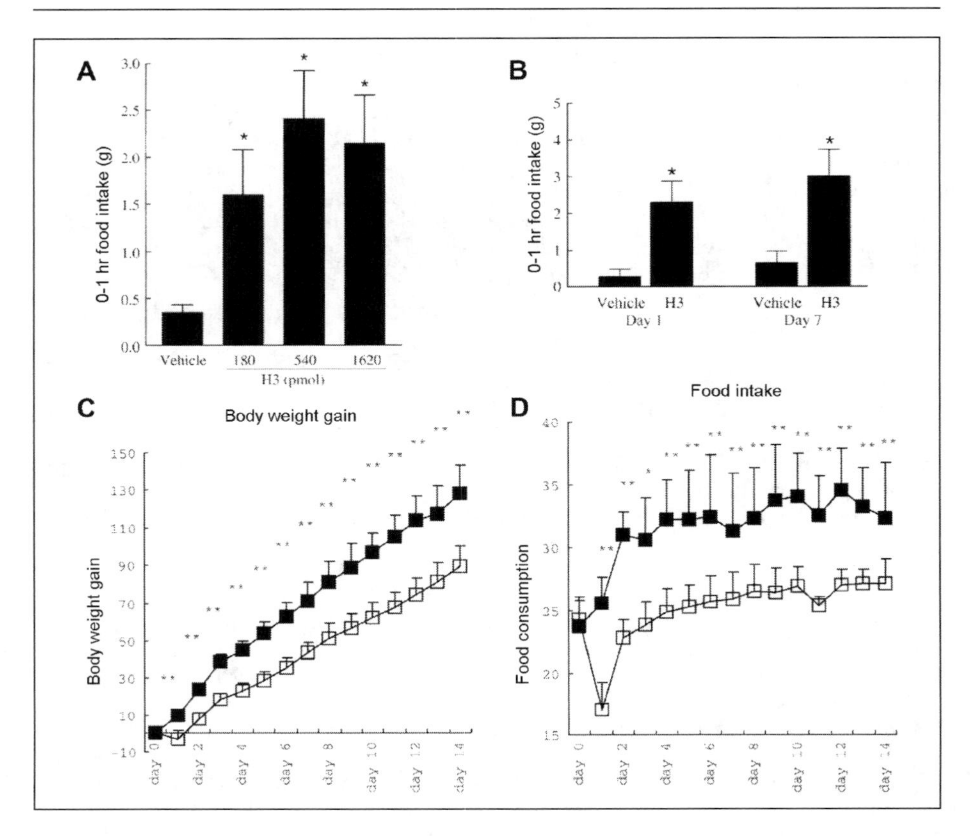

Figure 4. Effect of relaxin 3 on feeding in rats. (A) Effect of acute intra-paraventricular (iPVN) administration of human RLN3 (180–1620 pmol) on 1-h food intake in satiated male Wistar rats in the early light phase. *, $P < 0.05$ vs vehicle, n=10–12. (B) Effect of repeated iPVN administration of vehicle or hRLN3 (180 pmol/injection) in ad libitum fed rats on 1-h food intake in the early light phase on day 1 and day 7. *, $P > 0.05$ vs. vehicle, n = 8-11. From McGowan et al[98] with permission; copyright © 2006 Elsevier Ltd. (C) Cumulative body weight gain and (D) food consumption change during a chronic 14-day ICV infusion of vehicle or relaxin 3 (vehicle, relaxin 3, n = 7/7). Alzet minipumps were implanted on day 0. To account for the difference between vehicle and relaxin 3 injected groups, cumulative body weight gain from day 0 of the experiment was analyzed instead of body weight itself (C). Data are expressed as the mean ± SEM. Significance of changes: *, $P < 0.05$; **, $P < 0.01$, with respect to rats injected with vehicle. From Hida et al[99] with permission; ©2006 Taylor & Francis Group, LLC.

These data suggest that relaxin 3 may play a role in long-term control of food intake via actions at hypothalamic sites such as the PVN and arcuate nuclei and the lateral hypothalamus. Rats chronically administered relaxin 3 into the cerebrospinal fluid via osmotic minipumps (600 pmol/day for 2 weeks) also displayed significantly elevated food consumption, weight gain and epididymal fat mass.[99] Plasma leptin and insulin concentrations were also increased in these animals.

Effects on Reproduction

A relaxin-neutralizing monoclonal antibody, MCA1, has been shown to cause disruption of the birth process, when given both peripherally[100-102] and centrally.[87] When administered to pregnant female rats icv, MCA1 caused a significant decrease in the length of gestation,[103] but had no effect on the duration of straining or parturition, delivery interval, live birth rate, or newborn body weight,[103] whereas these processes were significantly disrupted during passive immunization by peripheral MCA1 administration.[101] The differential effects observed support the existence of central relaxin signaling that is somewhat independent of the peripheral relaxin system.

An in vivo experiment by Hsu et al[34] demonstrated that the soluble ectodomain of LGR7, designated 7BP, was a functional antagonist at LGR7 and subcutaneous administration (500 mg/day) for the final 4 days of gestation in pregnant mice led to a 27-h delay in parturition, further suggesting a functional role of central relaxin in the timing of pregnancy. This treatment also resulted in underdeveloped nipples, consistent with similar peripheral phenotypes reported for relaxin-KO[104] and LGR7-KO mice.[84] Interestingly however, the length of gestation is not altered in LGR7-KO mice,[84] suggesting a possible difference in relaxin function in mouse and rat brain (see below).

Effects on Learning and Memory

LGR7 is present in limbic regions known to participate in higher brain processes such as emotion, learning and memory, including the hippocampus, subiculum, entorhinal cortex and the basolateral complex of the amygdala (BLA). The BLA is very well characterized in regard to its efferent and afferent connections[105-107] and its involvement in emotion and modulation of fear memory.[108-111] Recent experiments using an aversively motivated, inhibitory avoidance paradigm tested the effects of relaxin on the activity of the BLA and consequent effects on memory retention for an aversive experience and demonstrated that infusions of relaxin into the BLA immediately after a footshock training session significantly impairs 48 h memory retention performance in a dose-dependent manner (Fig. 5).[65] This effect was specific to immediate processes of fear memory, (i.e., delayed infusions had no effect) and specific to the BLA. These results indicate therefore, that relaxin can alter early memory consolidation processes, although further studies are required to investigate the neurochemical mechanisms underlying this effect and whether endogenous relaxin can be shown to alter BLA function.

Brain regions such as the supramammillary nucleus, septum/nucleus of the diagonal band and hippocampus receive a strong relaxin 3-LI innervation and contain GPCR135; and are known to be involved in the generation of hippocampal theta rhythm and underlying memory processes. Recent

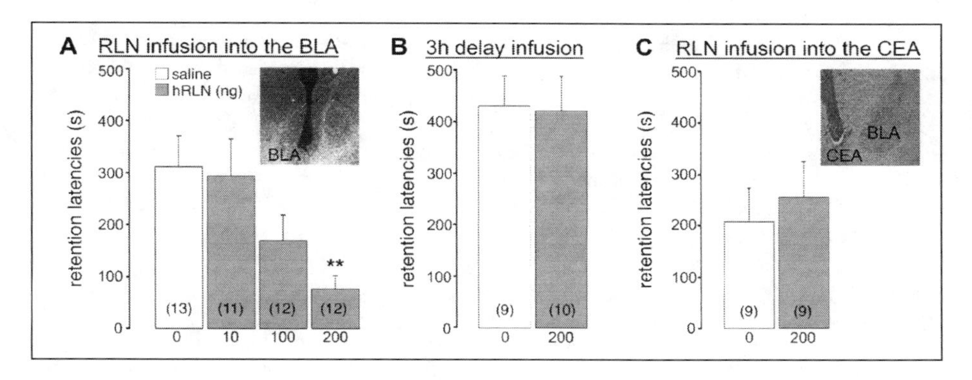

Figure 5. Effects of relaxin infusions into the amygdala on inhibitory avoidance retention latencies. Values in seconds (mean ± SEM; n = 9–13 rats per group) were measured during a 48-h posttraining inhibitory avoidance test. Results illustrate the effect of 0.2 μl infusions of: (A) saline or human relaxin (RLN; 10, 100, 200 ng) into the BLA immediately after training; (B) saline or RLN (200 ng) into the BLA, 3h after training; and (C) saline or RLN (200 ng) the CEA immediately after training. **, P < 0.01 compared with corresponding saline group. Modified from Figure 3 of Ma et al[65] with permission; ©2005 Blackwell Publishing.

studies by Nunez and others reported marked induction of hippocampal theta rhythm following electrical stimulation of the NI in anaesthetized rats.[112] Theta rhythm could also be induced by direct, electrical stimulation of the reticular pons or sensory stimulation (stroking the fur on the back), but lesions of the NI or injection of the $GABA_A$ agonist, muscimol, into the NI abolished theta rhythm evoked by reticular pons (oralis) stimulation.[112] On this basis it is believed that the NI may be an important relay station mediating excitatory transmission from the brainstem reticular formation to septohippocampal circuits related to theta rhythm. It will be of interest to explore whether relaxin 3 is involved in generation, modulation and/or maintenance of septohippocampal theta rhythm and subsequent effects on learning and memory in rats.

Relaxin 3 and stress. To date there have been very few studies investigating the role of relaxin 3 and GPCR135 in the CNS in relation to complex behavior, but their complementary anatomical distribution, associated with a network of NI GABAergic projections,[70] strongly suggests such functions for this peptide-receptor system, particularly relating to arousal and stress responses. In this regard, relaxin 3 nerve fibers and/or GPCR135 are observed in important areas such as the medial and lateral septum, paraventricular, supraoptic and dorsomedial hypothalamic and supramammillary nuclei, the central and medial amygdala and the periaqueductal gray, which mediate central autonomic changes in blood pressure, respiration, etc., associated with the stress response to anxiogenic stimuli.

Studies in our laboratory have revealed a rapid elevation in relaxin 3 mRNA levels in the NI following a repeat swim stress in rats (2 x 10 min sessions, 24-h apart; manuscript in preparation).[113] Elevation in NI relaxin 3 expression has also been reported after 6 h of restraint/water immersion stress in rats.[69] Further studies are now required to test the effects of other stressors and physiological stimuli on relaxin 3 expression in the NI and on peptide levels and/or release throughout relevant areas of the forebrain. These studies, along with further pharmacological experiments to activate or block specific GPCR135 populations in different brain areas should begin to elucidate the interactions occurring between relaxin 3 and other transmitter/peptide systems.

Future Directions and Developments in Central Relaxin Family Research

Various significant milestones in relaxin research, particularly the identification and characterization of the receptors responsive to relaxin, INSL3, INSL5 and the newly identified ancestral relaxin peptide—relaxin 3; and the generation of several relevant knock-out mouse strains, have allowed the development of a novel range of hypotheses related to central relaxin family peptide-receptor systems and provided the biological tools to explore them. Despite the slow pace of research on brain relaxin systems over the last decade, significant advancements have occurred, in particular, selective insights into the regional and cellular distribution of relaxin family peptides and their receptors in rat and mouse brain and more recent studies that strongly suggest functions of central relaxin, INSL3 and relaxin 3 signaling in homeostatic and higher-order cognitive processing and associated behaviors.

Future studies in this area should include further characterization of relaxin- and relaxin 3-producing neurons, in terms of co-expressed transmitters (glutamate, GABA) and neuropeptides or neurohormones and similarly studies of the precise phenotype of LGR7- LGR8- and GPCR135- expressing neurons in a range of important brain areas/circuits. In addition to further detailed neuroanatomical studies, it is important to begin exploring the functional effects of these central systems—for example, further investigate the underlying mechanisms of behavioral effects of relaxin in the BLA using LGR7 antagonist peptides once these are available and systematic investigations of the effect of exogenous relaxin 3 (or the R3/I5 chimera)[55] and/or a GPCR135 antagonist on the activity of GPCR135-rich regions of the brain and behavioral or neuroendocrine measures of stress responses, feeding and circadian processes. Aspects of relaxin neurobiology should also be further investigated in relaxin- and relaxin 3- and LGR7- and GPCR135- KO mice and additional relevant mouse strains and models. With respect to central INSL3/LGR8 systems, further detailed neuroanatomical studies are needed, particularly to identify the source of the endogenous ligand for central LGR8 and functional studies are needed to establish the predicted effects of LGR8 signaling on sensorimotor systems.

Another interesting aspect requiring further investigation is the role of relaxin/relaxin 3 and the abundant levels of LGR7/GPCR135 in adjacent regions of the olfactory bulb, accessory nuclei and cortex. These regions are involved in the associative and behavioral-level processes associated with olfactory information[114,115] and future investigations should aim to examine the effects of discrete injections of relaxin and relaxin 3 agonists and antagonists into these olfactory regions on olfactory discrimination tasks, such as inhibitory avoidance, utilizing perhaps cat odor as the aversive stimuli.[116] Olfactory processes are also related to the establishment of maternal behavior that occurs postpartum[117] and future studies should investigate the possible involvement of these brain systems in maternal and reproductive behavior. Furthermore, olfactory function in relaxin-family mutant and transgenic mice should be tested, e.g., testing for deficits in olfactory-related fear conditioning.

The generation of relaxin- and LGR7-KO mice has provided powerful tools for characterizing the effects of relaxin and LGR7 loss, although there are some obstacles with phenotyping such 'whole-of-life' mutant mice. The first is the presence of the mutation in all cells. Relaxin and LGR7 are expressed in the brain and various peripheral tissues and therefore it is difficult to assign a particular behavioral phenotype to a specific brain structure, or pathway, or even exclusively to the nervous system.[118] Tissue-specific 'conditional' transgenic technologies are able to overcome this problem, although these are yet to be applied to relaxin-family systems. Another potential problem is the loss of these genes during development and therefore observed changes in behavioral phenotype may be a consequence of alterations in both peripheral and central organ and tissue development. Furthermore, it is also possible that other genes may compensate during development for the loss of the target gene and therefore a behavioral phenotype may be attributable to effects of compensatory genes. Technology to create temporally-selective 'inducible' transgenic mice may overcome this problem, but again this is yet to be seen in the relaxin-field. However, studies to create viral vector-mediated knockdown or promotion of relaxin-family gene expression are being established and a recent study by Silvertown and colleagues[119] has demonstrated the effectiveness of such an approach. Indeed, the next decade should confirm important roles for relaxin and relaxin-related peptides in the central nervous system, associated with far more widespread functions than just reproduction.

References

1. Hisaw FL. Experimental relaxation of the pubic ligament of the guinea pig. Proc Soc Exp Biol Med 1926; 23:661-663.
2. Adham IM, Agoulnik AI. Insulin-like 3 signalling in testicular descent. Int J Androl 2004; 27:257-265.
3. Chassin D, Laurent A, Janneau JL et al. Cloning of a new member of the insulin gene superfamily (INSL4) expressed in human placenta. Genomics 1995; 29:465-470.
4. Conklin D, Lofton-Day CE, Haldeman BA et al. Identification of INSL5, a new member of the insulin superfamily. Genomics 1999; 60:50-56.
5. Hsu SY. Cloning of two novel mammalian paralogs of relaxin/insulin family proteins and their expression in testis and kidney. Mol Endocrinol 1999; 13:2163-2174.
6. Lok S, Johnston DS, Conklin D et al. Identification of INSL6, a new member of the insulin family that is expressed in the testis of the human and rat. Biol Reprod 2000; 62:1593-1599.
7. Kasik J, Muglia L, Stephan DA et al. Identification, chromosomal mapping and partial characterization of mouse Insl6: A new member of the insulin family. Endocrinology 2000; 141:458-461.
8. Bathgate RAD, Samuel CS, Burazin TCD et al. Human relaxin gene 3 (H3) and the equivalent mouse relaxin (M3) gene. Novel members of the relaxin peptide family. J Biol Chem 2002; 277:1148-1157.
9. Bathgate RAD, Hsueh AJ, Sherwood OD. Physiology and molecular biology of the relaxin peptide family. In: Neill JD, ed. Knobil and Neill's Physiology of Reproduction. New York: Academic Press, 2006:701-790.
10. Wilkinson TN, Speed TP, Tregear GW et al. Evolution of the relaxin-like peptide family. BMC Evol Biol 2005:5:14.
11. Adham IM, Burkhardt E, Benahmed M et al. Cloning of a cDNA for a novel insulin-like peptide of the testicular Leydig cells. J Biol Chem 1993; 268:26668-26672.

12. Pusch W, Balvers M, Ivell R. Molecular cloning and expression of the relaxin-like factor from the mouse testis. Endocrinology 1996; 137:3009-3013.
13. Burkhardt E, Adham IM, Brosig B et al. Structural organization of the porcine and human genes coding for a Leydig cell-specific insulin-like peptide (LEY I-L) and chromosomal localization of the human gene (INSL3). Genomics 1994; 20:13-19.
14. Koskimies P, Spiess AN, Lahti P et al. The mouse relaxin-like factor gene and its promoter are located within the 3' region of the JAK3 genomic sequence. FEBS Lett 1997; 419:186-190.
15. Zimmermann S, Schottler P, Engel W et al. Mouse Leydig insulin-like (Ley I-L) gene: Structure and expression during testis and ovary development. Mol Reprod Dev 1997; 47:30-38.
16. Spiess AN, Balvers M, Tena-Sempere M et al. Structure and expression of the rat relaxin-like factor (RLF) gene. Mol Reprod Dev 1999; 54:319-325.
17. Bullesbach EE, Gowan LK, Schwabe C et al. Isolation, purification and the sequence of relaxin from spiny dogfish (Squalus acanthias). Eur J Biochem 1986; 161:335-341.
18. Lu C, Walker WH, Sun J et al. Insulin-like peptide 6: Characterization of secretory status and posttranslational modifications. Endocrinology 2006; 147:5611-5623.
19. Liu C, Kuei C, Sutton S et al. INSL5 is a high affinity specific agonist for GPCR142 (GPR100). J Biol Chem 2005; 280:292-300.
20. Dun SL, Brailoiu E, Wang Y et al. Insulin-like peptide 5: Expression in the mouse brain and mobilization of calcium. Endocrinology 2006; 147:3243-3248.
21. Burazin TCD, Bathgate RAD, Macris M et al. Restricted, but abundant, expression of the novel rat gene-3 (R3) relaxin in the dorsal tegmental region of brain. J Neurochem 2002; 82:1553-1557.
22. Liu C, Eriste E, Sutton S et al. Identification of relaxin-3/INSL7 as an endogenous ligand for the orphan G-protein coupled receptor GPCR135. J Biol Chem 2003; 278:50754-50764.
23. Hsu SY. New insights into the evolution of the relaxin-LGR signaling system. Trends Endocrinol Metab 2003; 14:303-309.
24. Hudson P, Haley J, John M et al. Structure of a genomic clone encoding biologically active human relaxin. Nature 1983; 301:628-631.
25. Crawford RJ, Hudson P, Shine J et al. Two human relaxin genes are on chromosome 9. EMBO J 1984; 3:2341-2345.
26. Garibay-Tupas JL, Csiszar K, Fox M et al. Analysis of the 5'-upstream regions of the human relaxin H1 and H2 genes and their chromosomal localization on chromosome 9p24.1 by radiation hybrid and breakpoint mapping. J Mol Endocrinol 1999; 23:355-365.
27. Bathgate RAD, Scott D, Chung SWL et al. Searching the human genome database for novel relaxin-like peptides. Letts Pept Sci 2002; 8:129-132.
28. Fowler KJ, Clouston WM, Fournier RE et al. The relaxin gene is located on chromosome 19 in the mouse. FEBS Lett 1991; 292:183-186.
29. Schwabe C, Bullesbach EE. Relaxin: Structures, functions, promises and nonevolution. FASEB J 1994; 8:1152-1160.
30. Bryant-Greenwood GD, Schwabe C. Human relaxins: Chemistry and biology. Endocr Rev 1994; 15:5-26.
31. Gast MJ. Studies of luteal generation and processing of the high molecular weight relaxin precursor. Ann N Y Acad Sci 1982; 380:111-125.
32. Gast MJ. Characterization of preprorelaxin by tryptic digestion and inhibition of its conversion to prorelaxin by amino acid analogs. J Biol Chem 1983; 258:9001-9004.
33. Hudson P, Haley J, Cronk M et al. Molecular cloning and characterization of cDNA sequences coding for rat relaxin. Nature 1981; 291:127-131.
34. Hsu SY, Nakabayashi K, Nishi S et al. Activation of orphan receptors by the hormone relaxin. Science 2002; 295:671-674.
35. Tregear GW, Ivell R, Bathgate RAD et al. Proceedings of the Third International Conference on Relaxin and Related Peptides. Dordrecht: Kluwer Academic Publishers; 2001.
36. Bartsch O, Bartlick B, Ivell R. Relaxin signalling links tyrosine phosphorylation to phosphodiesterase and adenylyl cyclase activity. Mol Hum Reprod 2001; 7:799-809.
37. Palejwala S, Stein D, Wojtczuk A et al. Demonstration of a relaxin receptor and relaxin-stimulated tyrosine phosphorylation in human lower uterine segment fibroblasts. Endocrinology 1998; 139:1208-1212.
38. Hsu SY, Kudo M, Chen T et al. The three subfamilies of leucine-rich repeat-containing G protein-coupled receptors (LGR): identification of LGR6 and LGR7 and the signaling mechanism for LGR7. Mol Endocrinol 2000; 14:1257-1271.
39. Sudo S, Kumagai J, Nishi S et al. H3 relaxin is a specific ligand for LGR7 and activates the receptor by interacting with both the ectodomain and the exoloop 2. J Biol Chem 2003; 278:7855-7862.
40. Hsu SY, Nakabayashi K, Nishi S et al. Relaxin signaling in reproductive tissues. Mol Cell Endocrinol 2003; 202:165-170.

41. Bullesbach EE, Schwabe C. LGR8 signal activation by the relaxin-like factor. J Biol Chem 2005; 280:14586-14590.
42. Halls ML, Bathgate RA, Summers RJ. Relaxin family peptide receptors RXFP1 and RXFP2 modulate cAMP signaling by distinct mechanisms. Mol Pharmacol 2006; 70:214-226.
43. Kumagai J, Hsu SY, Matsumi H et al. INSL3/Leydig insulin-like peptide activates the LGR8 receptor important in testis descent. J Biol Chem 2002; 277:31283-31286.
44. Kawamura K, Kumagai J, Sudo S et al. Paracrine regulation of mammalian oocyte maturation and male germ cell survival. Proc Natl Acad Sci USA 2004; 101:7323-7328.
45. Anand-Ivell RJ, Relan V, Balvers M et al. Expression of the insulin-like peptide 3 (INSL3) hormone-receptor (LGR8) system in the testis. Biol Reprod 2006; 74:945-953.
46. Zimmermann S, Steding G, Emmen JM et al. Targeted disruption of the Insl3 gene causes bilateral cryptorchidism. Mol Endocrinol 1999; 13:681-691.
47. Feng S, Cortessis VK, Hwang A et al. Mutation analysis of INSL3 and GREAT/LGR8 genes in familial cryptorchidism. Urology 2004; 64:1032-1036.
48. Gorlov IP, Kamat A, Bogatcheva NV et al. Mutations of the GREAT gene cause cryptorchidism. Hum Mol Genet 2002; 11:2309-2318.
49. Shen PJ, Fu P, Phelan KD et al. Restricted expression of LGR8 in intralaminar thalamic nuclei of rat brain suggests a role in sensorimotor systems. Ann N Y Acad Sci 2005; 1041:510-515.
50. Sedaghat K, Shen P-J, Allbutt H et al. Localization, lesion and functional studies of the INSL3 receptor, LGR8, in brain: Further evidence for role in sensorimotor pathways. Soc Neurosci Abstr 2005: P725.711.
51. Sedaghat K, Shen P-J, Bathgate RAD et al. Leucine-rich repeat-containing G-protein-coupled receptor 8 (LGR8) in rat sensorimotor-basal ganglia circuitry. Soc Neurosci Abstr 2006:P450.24.
52. Liu C, Chen J, Sutton S et al. Identification of relaxin-3/INSL7 as a ligand for GPCR142. J Biol Chem 2003; 278:50765-50770.
53. Matsumoto M, Kamohara M, Sugimoto T et al. The novel G-protein coupled receptor SALPR shares sequence similarity with somatostatin and angiotensin receptors. Gene 2000; 248:183-189.
54. Sutton SW, Bonaventure P, Kuei C et al. G-protein-coupled receptor (GPCR)-142 does not contribute to relaxin-3 binding in the mouse brain: Further support that relaxin-3 is the physiological ligand for GPCR135. Neuroendocrinology 2006; 82:139-150.
55. Liu C, Chen J, Kuei C et al. Relaxin-3/insulin-like peptide 5 chimeric peptide, a selective ligand for G protein-coupled receptor (GPCR)135 and GPCR142 over leucine-rich repeat-containing G protein-coupled receptor 7. Mol Pharmacol 2005; 67:231-240.
56. Sherwood OD. Relaxin's physiological roles and other diverse actions. Endocr Rev 2004; 25:205-234.
57. Gunnersen JM, Crawford RJ, Tregear GW. Expression of the relaxin gene in rat tissues. Mol Cell Endocrinol 1995; 110:55-64.
58. Samuel CS, Tian H, Zhao L et al. Relaxin is a key mediator of prostate growth and male reproductive tract development. Lab Invest 2003; 83:1055-1067.
59. Ivell R, Hartung S. The molecular basis of cryptorchidism. Mol Hum Reprod 2003; 9:175-181.
60. Ferlin A, Simonato M, Bartoloni L et al. The INSL3-LGR8/GREAT ligand-receptor pair in human cryptorchidism. J Clin Endocrinol Metab 2003; 88:4273-4279.
61. Roh J, Virtanen H, Kumagai J et al. Lack of LGR8 gene mutation in Finnish patients with a family history of cryptorchidism. Reprod Biomed Online 2003; 7:400-406.
62. Ferlin A, Bogatcheva NV, Gianesello L et al. Insulin-like factor 3 gene mutations in testicular dysgenesis syndrome: Clinical and functional characterization. Mol Hum Reprod 2006; 12:401-406.
63. Osheroff PL, Ho W-H. Expression of relaxin mRNA and relaxin receptors in postnatal and adult rat brains and hearts. J Biol Chem 1993; 268:15193-15199.
64. Burazin TCD, Davern P, McKinley MJ et al. Identification of relaxin and relaxin responsive cells in the rat brain. In: Tregear GW, Ivell R, Bathgate RAD, Wade JD, ed. Proceedings of the Third International Conference on Relaxin and Related Peptides. Dordrecht: Kluwer Academic Publishers, 2001:209-214.
65. Ma S, Roozendaal B, Burazin TCD et al. Relaxin receptor activation in the basolateral amygdala impairs memory consolidation. Eur J Neurosci 2005; 22:2117-2122.
66. Allen Institute for Brain Science. Allen Brain Atlas [Website]. Available at: http://www.brain-map.org. Accessed October, 2006.
67. Hokfelt T, Broberger C, Xu ZQ et al. Neuropeptides—an overview. Neuropharmacology 2000; 39:1337-1356.
68. Ludwig M, Leng G. Dendritic peptide release and peptide-dependent behaviours. Nat Rev Neurosci 2006; 7:126-136.
69. Tanaka M, Iijima N, Miyamoto Y et al. Neurons expressing relaxin 3/INSL 7 in the nucleus incertus respond to stress. Eur J Neurosci 2005; 21:1659-1670.

70. Ma S, Bonaventure P, Ferraro T et al. Relaxin-3 in GABA projection neurons of nucleus incertus suggests widespread influence on forebrain circuits via G-protein-coupled receptor-135 in the rat. Neuroscience 2007; 144:165-190.
71. Papez J. Comparative Neurology. New York, 1929.
72. Chatfield PO, Lyman CP. An unusual structure in the floor of the fourth ventricle of the golden hamster, Mesocricetus auratus. J Comp Neurol 1954; 101:225-235.
73. Meesen H, Olszewski J. A Cytoarchitectonic Atlas of the Rhombencephalon of the Rabbit. Basel, 1949.
74. Paxinos G, Watson C. The Rat Brain in Stereotaxic Coordinates. New York: Academic Press, 1986.
75. Jennes L, Stumpf WE, Kalivas PW. Neurotensin: Topographical distribution in rat brain by immunohistochemistry. J Comp Neurol 1982; 210:211-224.
76. Morest DK. Connexions of the dorsal tegmental nucleus in rat and rabbit. J Anat 1961; 95:229-246.
77. Bittencourt JC, Sawchenko PE. Do centrally administered neuropeptides access cognate receptors? An analysis in the central corticotropin-releasing factor system. J Neurosci 2000; 20:1142-1156.
78. Goto M, Swanson LW, Canteras NS. Connections of the nucleus incertus. J Comp Neurol 2001; 438:86-122.
79. Olucha-Bordonau FE, Teruel V, Barcia-Gonzalez J et al. Cytoarchitecture and efferent projections of the nucleus incertus of the rat. J Comp Neurol 2003; 464:62-97.
80. Bathgate RAD, Ivell R, Sanborn BM et al. International Union of Pharmacology LVII: Recommendations for the nomenclature of receptors for relaxin family peptides. Pharmacol Rev 2006; 58:7-31.
81. Osheroff PL, Phillips HS. Autoradiographic localization of relaxin binding sites in rat brain. Proc Natl Acad Sci USA 1991; 88:6413-6417.
82. Ma S, Shen PJ, Burazin TCD et al. Comparative localization of leucine-rich repeat-containing G-protein-coupled receptor-7 (RXFP1) mRNA and [(33)P]-relaxin binding sites in rat brain: Restricted somatic coexpression a clue to relaxin action? Neuroscience 2006; 141:329-344.
83. Sutton SW, Bonaventure P, Kuei C et al. Distribution of G-protein-coupled receptor (GPCR)135 binding sites and receptor mRNA in the rat brain suggests a role for relaxin-3 in neuroendocrine and sensory processing. Neuroendocrinology 2004; 80:298-307.
84. Krajnc-Franken MA, Van Disseldorp AJ, Koenders JE et al. Impaired nipple development and parturition in LGR7 knockout mice. Mol Cell Biol 2004; 24:687-696.
85. Piccenna L, Shen P-J, Ma S et al. Localization of LGR7 gene expression in adult mouse brain using LGR7 knock-out/LacZ knock-in mice: Correlation with LGR7 mRNA distribution. Ann N Y Acad Sci 2005; 1041:197-204.
86. Sinnayah P, Burns P, Wade JD et al. Water drinking in rats resulting from intravenous relaxin and its modification by other dipsogenic factors. Endocrinology 1999; 140:5082-5086.
87. Summerlee AJ, Hornsby DJ, Ramsey DG. The dipsogenic effects of rat relaxin: The effect of photoperiod and the potential role of relaxin on drinking in pregnancy. Endocrinology 1998; 139:2322-2328.
88. Gross PM. Circumventricular organ capillaries. Prog Brain Res 1992; 91:219-233.
89. Sunn N, McKinley MJ, Oldfield BJ. Identification of efferent neural pathways from the lamina terminalis activated by blood-borne relaxin. J Neuroendocrinol 2001; 13:432-437.
90. Bathgate RAD, Lin F, Hanson NF et al. Relaxin-3: Improved synthesis strategy and demonstration of its high-affinity interaction with the relaxin receptor LGR7 both in vitro and in vivo. Biochemistry 2006; 45:1043-1053.
91. Thornton SM, Fitzsimons JT. The effects of centrally administered porcine relaxin on drinking behaviour in male and female rats. J Neuroendocrinol 1995; 7:165-169.
92. O'Byrne KT, Eltringham L, Summerlee AJ. Central inhibitory effects of relaxin on the milk ejection reflex of the rat depends upon the site of injection into the cerebroventricular system. Brain Res 1987; 405:80-83.
93. Parry LJ, Poterski RS, Summerlee AJ. Effects of relaxin on blood pressure and the release of vasopressin and oxytocin in anesthetized rats during pregnancy and lactation. Biol Reprod 1994; 50:622-628.
94. Heine PA, Di S, Ross LR anderson LL et al. Relaxin-induced expression of Fos in the forebrain of the late pregnant rat. Neuroendocrinology 1997; 66:38-46.
95. McKinley MJ, Burns P, Colvill LM et al. Distribution of Fos immunoreactivity in the lamina terminalis and hypothalamus induced by centrally administered relaxin in conscious rats. J Neuroendocrinol 1997; 9:431-437.
96. Cronin MJ, Malaska T. Characterization of relaxin-stimulated cyclic AMP in cultured rat anterior pituitary cells: Influence of dopamine, somatostatin and gender. J Mol Endocrinol 1989; 3:175-182.
97. McGowan BM, Stanley SA, Smith KL et al. Central relaxin-3 administration causes hyperphagia in male Wistar rats. Endocrinology 2005; 146:3295-3300.
98. McGowan BM, Stanley SA, Smith KL et al. Effects of acute and chronic relaxin-3 on food intake and energy expenditure in rats. Regul Pept 2006; 136:72-77.

99. Hida T, Takahashi E, Shikata K et al. Chronic intracerebroventricular administration of relaxin-3 increases body weight in rats. J Recept Signal Transduct Res 2006; 26:147-158.
100. Hwang JJ, Sherwood OD. Monoclonal antibodies specific for rat relaxin. III. Passive immunization with monoclonal antibodies throughout the second half of pregnancy reduces cervical growth and extensibility in intact rats. Endocrinology 1988; 123:2486-2490.
101. Guico-Lamm ML, Sherwood OD. Monoclonal antibodies specific for rat relaxin. II. Passive immunization with monoclonal antibodies throughout the second half of pregnancy disrupts birth in intact rats. Endocrinology 1988; 123:2479-2485.
102. Kuenzi MJ, Sherwood OD. Monoclonal antibodies specific for rat relaxin. VII. Passive immunization with monoclonal antibodies throughout the second half of pregnancy prevents development of normal mammary nipple morphology and function in rats. Endocrinology 1992; 131:1841-1847.
103. Summerlee AJ, Ramsey DG, Poterski RS. Neutralization of relaxin within the brain affects the timing of birth in rats. Endocrinology 1998; 139:479-484.
104. Zhao L, Roche PJ, Gunnersen JM et al. Mice without a functional relaxin gene are unable to deliver milk to their pups. Endocrinology 1999; 140:445-453.
105. Price JL. Comparative aspects of amygdala connectivity. Ann N Y Acad Sci 2003; 985:50-58.
106. Sah P, Faber ES, Lopez De Armentia M et al. The amygdaloid complex: Anatomy and physiology. Physiol Rev 2003; 83:803-834.
107. Pitkanen A. Connectivity of the rat amygdaloid complex. In: Aggleton JP, ed. The Amygdala: A Functional Analysis. Oxford, UK: Oxford University Press, 2000:31-115.
108. Cardinal RN, Parkinson JA, Hall J et al. Emotion and motivation: The role of the amygdala, ventral striatum and prefrontal cortex. Neurosci Biobehav Rev 2002; 26:321-352.
109. Hamann SB, Ely TD, Grafton ST et al. Amygdala activity related to enhanced memory for pleasant and aversive stimuli. Nat Neurosci 1999; 2:289-293.
110. McGaugh JL, McIntyre CK, Power AE. Amygdala modulation of memory consolidation: Interaction with other brain systems. Neurobiol Learn Mem 2002; 78:539-552.
111. McGaugh JL. The amygdala modulates the consolidation of memories of emotionally arousing experiences. Annu Rev Neurosci 2004; 27:1-28.
112. Nunez A, Cervera-Ferri A, Olucha-Bordonau F et al. Nucleus incertus contribution to hippocampal theta rhythm generation. Eur J Neurosci 2006; 23:2731-2738.
113. Banerjee A, Ma S, Ortinau S et al. Relaxin-3 neurons in the nucleus incertus: Projection patterns, response to swim stress and RLX3 neuronal signaling. Soc Neurosci Abstr 2005; 35:59.57.
114. Hamann S. Nosing in on the emotional brain. Nat Neurosci 2003; 6:106-108.
115. Schoenbaum G, Chiba AA, Gallagher M. Neural encoding in orbitofrontal cortex and basolateral amygdala during olfactory discrimination learning. J Neurosci 1999; 19:1876-1884.
116. Dielenberg RA, McGregor IS. Defensive behavior in rats towards predatory odors: A review. Neurosci Biobehav Rev 2001; 25:597-609.
117. Levy F, Keller M, Poindron P. Olfactory regulation of maternal behavior in mammals. Horm Behav 2004; 46:284-302.
118. Crawley JN. Behavioral phenotyping of transgenic and knockout mice: Experimental design and evaluation of general health, sensory functions, motor abilities and specific behavioral tests. Brain Res 1999; 835:18-26.
119. Silvertown JD, Geddes BJ, Summerlee AJ. Adenovirus-mediated expression of human prorelaxin promotes the invasive potential of canine mammary cancer cells. Endocrinology 2003; 144:3683-3691.
120. Bell GI, Pictet RL, Rutter WJ et al. Sequence of the human insulin gene. Nature 1980; 284:26-32.
121. Jansen M, van Schaik FM, Ricker AT et al. Sequence of cDNA encoding human insulin-like growth factor I precursor. Nature 1983; 306:609-611.
122. Bell GI, Merryweather JP, Sanchez-Pescador R et al. Sequence of a cDNA clone encoding human preproinsulin-like growth factor II. Nature 1984; 310:775-777.

Index

F

G

H

I

K

L

M

N

O

P

R